Once Upon Einstein

Once Upon Einstein

Thibault Damour

translated by Eric Novak

A K Peters, Ltd.
Wellesley, Massachusetts

Editorial, Sales, and Customer Service Office

A K Peters, Ltd.
888 Worcester Street, Suite 230
Wellesley, MA 02482
www.akpeters.com

Originally published in French.
Si Einstein m'était conté, copyright © Le Cherche Midi Editeur, 2005.
English translation by Eric Novak.

Library of Congress Cataloging-in-Publication Data

Damour, Thibault.
    [Si Einstein m'était conté. English]
    Once upon einstein / Thibault Damour ; translated by Eric Novak.
      p. cm.
    Includes bibliographical references.
    ISBN 13: 978-1-56881-289-2 (alk. paper)
    ISBN 10: 1-56881-289-2 (alk. paper)
      1. Einstein, Albert, 1879-1955. 2. Physicists--Germany--History--20th century--Biography. 3. Physicists--United States--History--20th century--Biography. 4. Physics--History--20th century. I. Title.

QC16.E5D35813 2006
530'.092--dc22
[B]
                                                                          2005056604

Cover photograph: Albert Einstein ca. 1930. Credit Rue des Archives / TAL.

Printed in the United States of America
10 09 08 07 06   10 9 8 7 6 5 4 3 2 1

To Thierry, who showed me the path,
to Hélène, who shared the path,
to Fax and Ulysses, who walked with me along the path.

Contents

Once Upon a Time . . .

Once upon a time, there was a five-year-old boy, lying ill in bed. To help him pass the time, his father brought him a compass. The child was struck by the strangeness of the magnetic needle. In place of what he had come to expect, the usual behavior of mundane objects, the needle always pointed in the same direction, however he might rotate the casing. How could something so extraordinary be possible? What was the hidden cause of this strange behavior?

The boy's name was Albert Einstein. For his entire life, he remembered the profound astonishment brought to him by the compass. This event was the first manifestation of what would guide him throughout his career: an insatiable *curiosity*, a desire to grasp the *hidden order beneath the surface appearance of things*. As he stated it:

> I know perfectly well that I myself have no special talents. It was curiosity, obsession, and sheer perseverance that brought me to my ideas.

It has been 100 years since the "miraculous year," 1905, during which Einstein laid the foundations of two of the crowning scientific theories of the twentieth century: special relativity and quantum mechanics. It has been 90 years since he completed the theory of general relativity, and 50 years since his death. Still the name Albert Einstein continues to fascinate the scientist as well as the man on the street. His life was rich and complex, but it was guided by this single, simple desire: to understand the hidden order of the universe. His success in this quest was extraordinary. Even if the second part of his life brought fewer concrete results than the first, the great ambition of his curiosity can be measured today by the fact that the questions which he posed then have only recently begun to be answered.

This book is not a biography of Einstein. We shall hardly speak of Einstein the husband, the father, the musician, the pacifist, or the Zionist. We shall not discuss his youth in Munich, his studies in Zurich, his difficulty in finding a job, his university career, his social life in the vibrant Berlin of

the 1920s, the letter that he wrote to Franklin Delano Roosevelt mentioning the possibility of building a nuclear bomb, nor his reclusive life in the
small college town of Princeton. This book is even less a course on Einstein's
theories, or a review of modern physics. Rather, this book will try to put
the reader in Einstein's place, and will encourage the reader to share some of
those particular moments when Einstein succeeded in "lifting a corner of the
great veil"—those times when he understood some part of the hidden order
of the universe. For someone like Einstein, these moments made up the very
essence of his life. We may quote him from his own "obituary," a set of autobiographical notes that he produced late in life, at the request of the editor
Paul Arthur Schilpp. After retracing the development of his ideas, with very
little mention of any biographical details, he continues:

> "Is this supposed to be an obituary?" the astonished reader will likely ask. I
> would like to reply: essentially yes. For the essential in the being of a man
> of my type lies precisely in *what* he thinks and *how* he thinks, not in what
> he does or suffers. Consequently, the obituary can limit itself in the main
> to the communication of thoughts which have played a considerable rôle in
> my endeavors.

We hope that the reader will be able to taste a bit of this "joy of thought,"
which Einstein experienced until his final day.

A small practical note before we call on Einstein to live and think in front
of us. To keep the text unencumbered, all of the technical details, as well as
some bibliographic references, have been collected as notes at the end of the
book. The reading of these notes is absolutely unnecessary for comprehension
of the text. They are simply there for the curious reader who may wish to
know more.

Acknowledgments

This book could not have existed without the varied contributions of many people, in addition to those already mentioned in the dedication. I would first like to thank John Wheeler who agreed to welcome (in 1974) the young and inexperienced *normalien* that I was at Princeton University, and Remo Ruffini with whom I began to understand physics. A warm (posthumous) thank you to Helen Dukas for an unforgettable afternoon spent taking tea, and remembering the great man, in his house on 112 Mercer Street. I would particularly like to thank Charlie Misner for his memories of Hugh Everett and Einstein's last seminar.

I have benefited, through the years, from the aid, suggestions, or memories of many friends and colleagues, in particular: Peter Bergmann, Curtis Callan, Yvonne Choquet-Bruhat, Stanley Deser, Bryce and Cécile DeWitt, Jean Eisenstaedt, David Gross, Marc Henneaux, Florent Hillaire, Hermann Nicolai, Jean Osty, Patrick Pierrel, Sasha Polyakov, Nathan Rosen, Wolfgang Schleich, Christophe Soulé, John Stachel, Norbert Straumann, and Thierry Thomas. Special thanks are due to my friend Jean Orizet, without whom this book would never have seen the light.

I warmly thank Cécile Cheikhchoukh for the friendliness and efficacy with which she contributed to the realization of this book, and Marie-Claude Vergne for the care she took in creating the figures.

Finally, I would like to thank Klaus Peters for suggesting the translation of this book within A K Peters, Ltd., and Eric Novak for the care and accuracy with which he translated the text into English.

Einstein at the patent office for intellectual property in Bern, 1902. (*Credit akg-images.*)

1

The Question of Time

*It turned out, surprisingly, that it was only necessary to define the time
concept precisely enough to overcome the difficulty.*
—Einstein

The Eagle and the Sparrow

Bern, Switzerland, May 1905

It was an exceptionally warm and beautiful May Sunday.[1] Finally, a nice
stretch of free time, after working all week at the Bern Patent Office! Since
that morning, Albert Einstein had been enjoying the opportunity to spend
an entire day thinking about the scientific problem that had been obsessing
him for years. Not that he thought of it ceaselessly—far from it! Indeed his
thoughts had been quite busy with other research for the first few months of
1905.

On March 14, he had celebrated his twenty-sixth birthday by putting the
finishing touches on a paper that he himself called "very revolutionary," in
which he had questioned the wave nature of light. At the end of March, and
during April, he had written what was to become his doctoral thesis: a new
calculation of the size of atoms and molecules. In early May, he had obtained
the basic laws governing the erratic motion of tiny grains of pollen suspended
in liquid. This motion reflected, at a macroscopic level, the molecular motion
connected to heat.[2] Since Wednesday, May 10, when he had sent this latest
work to a scientific journal, Einstein had re-immersed himself, during his rare
free moments, to the problem he had been considering for ten years.

Ten years of thoughtful consideration for a young man of 26 years? Indeed, for when Einstein was 16 years old, in 1895, he had suddenly been
struck by a question: could one chase after a ray of light, and catch up with

it? It had long been known[3] that light traveled at a finite speed, around 300,000 kilometers per second. Moreover, according to the ideas about space and time that everyone had agreed on since the founding work of Galileo, Descartes, and Newton, there was nothing to stop one from attaining or even surpassing, in principle, this speed. Likewise, nothing could stop the young Einstein from imagining an observer moving at the speed of light. But what would such an observer, straddling a light beam, see? The young Albert remembered the enthusiastic explanations given by his uncle Jakob of the nature of light, as described by the theoretical calculations of James Clerk Maxwell in the mid-nineteenth century, and established, in 1887, by the experiments of Heinrich Hertz. Light was an *electromagnetic wave*: a traveling oscillation in which an electric field played "leap-frog" with a magnetic field. At each point in space, the strength of each field oscillated regularly between positive and negative values, like the height of water in a normal water wave, but with the two different fields staggered with respect to each other, so that, when the electric field has its maximum amplitude, the magnetic field has an amplitude of zero, and vice versa. Moreover, just as the "frogs" move forward in a game of leap-frog (here with an infinite number of electric frogs alternating with an infinite number of magnetic frogs), this doubly oscillating "game" travels through space at a speed of 300,000 km/s, like a wave on the surface of the sea. It would thus seem that an observer who was moving at the same speed as an electromagnetic wave should see, like a surfer riding a wave on the ocean, not an oscillation, but a configuration of electric and magnetic fields which, although repeating itself periodically in space, remains the same as time passes. However, Maxwell's equations, which successfully describe the behavior of electromagnetic fields, do not admit any such solutions. Faced with this contradiction, the young Albert intuitively deduced that it was probably not possible to travel at the speed of light.

This youthful "thought experiment" remained in the back of Einstein's mind during his advanced scientific studies at the Federal Polytechnic Institute in Zurich, Switzerland. This school was renowned throughout Europe for the quality of its teaching and the scientific prestige of its professors. For example, the physics curriculum was directed by Heinrich Weber, the author of some important work on the heat capacity of various solids (such as diamond). Einstein had hoped that Weber's course would cover the latest developments in the theory of electromagnetic fields. But such was not the case. Weber's course went no further than what Einstein already knew; it was a simple introduction to Maxwell's established theory of electromagnetism,[4] and a brief glimpse of Hertz's experimental results. As a result, Einstein "ditched" Weber's formal lectures in order to study, alone, some more recent books on electromagnetism, as well as some of the original articles (including those of Hertz). He also spent much of his time in the Polytechnic's modern and well-equipped experimental laboratory, going so far as to propose an experiment to measure the speed of the Earth with respect to the *ether* (this proposal, however, was rejected by Weber, who didn't care much for this student who, although quite intelligent, had a self-confidence bordering on insolence). Ether was the name given to the medium through which light, as an electromagnetic field, propagated. For the physicists of the nineteenth century, the existence of a propagating "medium" for light and electromagnetic fields seemed an absolute necessity, in the same way that air is needed to carry sound.

After his graduation in 1900 from the Polytechnic, Einstein had continued to reflect on the possibility of measuring the speed of the Earth with respect to the ether. He had read an important paper by the Dutch scientist Hendrik Lorentz which mentioned the failure of all such attempted experiments, and which showed that Maxwell's theory had curious properties once one considered it, for example, from Earth's reference frame, moving through the ether, rather than from the reference frame at rest with respect to the ether. These properties tended to make it extremely difficult, if not impossible, for terrestrial observers to detect their motion with respect to the ether.

These, then, were the problems which Einstein promised himself to attack anew on this beautiful May Sunday in 1905: *What would an observer see who tries to catch up with a ray of light (or an electromagnetic wave)? Could an observer being dragged along with the Earth detect his motion with respect to the "ether"?*

Unfortunately, after a morning spent reflecting in the Besenscheuerweg apartment to which he had recently moved, Einstein could not see any clearer into these problems. In spite of his powers of concentration, it was difficult for him to calmly reflect in this small apartment, where his wife, Mileva, busied herself with housework and small Hans Albert, who had just celebrated his first birthday, was beginning to walk and was wandering about. Einstein decided that the best way to make progress in his reflections would be to pay a visit to his friend Michele Angelo Besso and to discuss matters with him while strolling through the hills outside Bern. Indeed, for a week, since Einstein had left his Kramgasse apartment in the historic center and installed himself in the Mattenhof quarter at Bern's periphery, he and Besso had practically been neighbors.

Michele was his oldest friend in Bern. They had first met in Zurich at some musical *soirées*, while Einstein was still preparing to enter the Polytechnic, from which Michele had recently graduated. From their first meeting, they knew that they would be life-long friends. They had so much in common. The love of music, of course, which was the direct cause for their encounter, but above all an intellectual curiosity knowing no limits. Science, literature, art, philosophy—everything was included. Moreover, Albert had introduced Michele to the woman who quickly became his wife: Anna Barbara Winteler, his landlord's daughter. The landlord, "Papa" Winteler, was a teacher at Aarau's cantonal school, where Albert had finished his secondary education and prepared himself for the Polytechnic's entry exams. For over a year Michele had joined Albert as a technical expert at the Federal Patent Office in Bern. They were thus able to see each other every day at the office, and they regularly accompanied each other on the return home. Since they had become neighbors, they even shared the trip to work. These trips were host to a series of animated discussions on a countless variety of subjects. When the discussions were focused on those scientific subjects which Albert worked on, Michele compared himself to a sparrow, carried to the heights by an eagle. The sparrow could not fly so high alone, but sometimes, while up there, it so happened that the sparrow could flutter about for a moment above the eagle.

Albert was reinvigorated by the decision to visit Michele. He bounded down the small stairway of his apartment. It was just after noon. Michele had no doubt also finished his lunch and would be happy to embark on a *passeggiata scientifica*. Soon the two friends were quickly climbing the hill of Gurten, with its magnificent view of Bern. Let's try to imagine their lively dialogue.

A.E.: I am sure that good old Galileo had it right, "the motion is as nothing."

M.B.: Meaning that one cannot detect one's own motion through space, as long as one moves at a constant speed, and in a straight line.

A.E.: Yes. And that should apply not only to mechanical properties . . .

M.B.: . . . to the butterflies fluttering in the cabin of a moving ship, which Galileo spoke of, . . .

A.E.: . . . but also to electromagnetic properties, and more generally to all the laws of physics. This is why I think that the concept of ether which they beat into us has no meaning.

M.B.: You're saying that Newton's absolute space, identified as the ether, doesn't exist. There should only be a relative space, as Leibniz suggested.

A.E.: Yes, something like that. It is too simple and elegant not to be true. And this would take care, in one fell swoop, of what Lorentz tries to demonstrate through the accumulation of various hypotheses and inelegant calculations, without ever really succeeding. However, I have a problem with the speed of light. If I assume that everything is relative, then a moving observer, like the fellow running after a light beam that I told you about, must not only never catch up with it, but must always see it moving at the same speed.

M.B.: But that's contrary to the law of addition of speeds! Let's go back to Galileo and his butterflies fluttering in the ship's cabin. The speed of a butterfly with respect to the shore is the sum[5] of its speed with respect to the ship and of the speed of the ship with respect to the shore.

A.E.: Yes, this is what stumps me. Nevertheless, I am convinced that the speed of light must always be the same: 300,000 kilometers per second, independently of the observer who measures it.

M.B.: I see your problem. It's a real contradiction. You would like that in adding an extra speed to 300,000 kilometers per second, one still obtains 300,000 kilometers per second! You may as well ask that 1 plus 1 equals 1, rather than 2! Note ... it's not logically excluded. You just have to modify the meaning of the symbol $+$. I remember Hurwitz's mathematics class at the Polytechnic. He gave examples of ways to combine numbers, modified definitions of addition, where one could have $1 + 1 = 1$.

A.E.: Yes, but we're speaking of good old numbers and of good old addition. After all, we're only adding kilometers per second.

M.B.: Okay, good. Then let's peel away at the question. What does "kilometer" mean? What does "second" mean?

A.E.: Wait ... that's it! I think I've got it ... look at the clock tower, down there in the center of town. If we had binoculars, we could see the time. But it would not be *our* time. We would need to account for the time taken for the light to come from the clock to us. I think that this will modify the notion of time for a moving observer. Thank you, Michele! I am sure that it will work now! Tonight, I shall work it out in detail.

M.B.: The sparrow doesn't understand what the piercing gaze of the eagle has seen, but he is happy to have helped, even a little. Meanwhile, *sol lucet omnibus*;[6] let's take advantage of this magnificent day!

That very night, Einstein verified that, in fact, everything worked. The next morning, when he saw Besso, he thanked him again for putting him on the right track. Six weeks later, in the last days of June, having put to good use those few moments of free time after (or even during) his hours at the office and on Sundays (and left over from his obligations as a husband and father), he sent the *Annalen der Physik* the founding article on the theory of relativity. This article contains no references to previous scientific works, but finishes with the sentence: "In conclusion, let me note that my friend and colleague M. Besso steadfastly stood by me in my work on the problem discussed here, and that I am indebted to him for several valuable suggestions. Bern, June 1905."

This short article of Einstein's is one of the most important scientific articles of the twentieth century. It is also one of the most "beautiful." It has an axiomatic perfection worthy of the treatments of Euclidean geometry that Einstein had so appreciated as a child. The logic flows without any apparent effort, like some of the most beautiful passages in Mozart. I strongly advise every young (or not so young) mind interested in reliving one of the great moments of human thought to read it themselves.[7] We shall sketch the contents and logic of it below.

Space and Time before Einstein

These gentlemen maintain therefore, that Space is a real absolute being. But this involves them in great difficulties.
—Leibniz

To understand how deeply Einstein's article from June 1905 changed the ages-old concepts of space and time, let's back up a bit. We shall not try to retrace the slow and tortuous development of these concepts, starting from the creation of geometry by the Greeks, moving through the reassuring vision of human reality held in the closed, medieval world, until finally this womb-like world was shattered and replaced by an infinite universe devoid of any concrete features.[8] We shall start from the concepts of "absolute" time and space, as they were crystallized in the *Mathematical Principles of Natural Philosophy*,[9] written by Isaac Newton in 1686. Let's revisit the famous commentary (or *scholium*) that Newton added, after having introduced the conceptual framework of his treatise:

> Hitherto I have laid down the definitions of such words as are less known, and explained the sense in which I would have them to be understood in the following discourse. I do not define *time, space, place* and *motion*, as being well known to all. Only I must observe, that the vulgar conceive those quantities under no other notions but from the relation they bear to sensible objects. And thence arise certain prejudices, for the removing of which, it will be convenient to distinguish them into *absolute* and *relative*, *true* and *apparent*, *mathematical* and *common*.
>
> I. Absolute, true, and mathematical time, of itself, and from its own nature flows equably without regard to anything external, and by another name is called *duration*; relative, apparent, and common time, is some sensible and external (whether accurate or unequable) measure of duration by the means of motion, which is commonly used instead of true time: such as an *hour*, a *day*, a *month*, a *year*.
>
> II. Absolute space, in its own nature, without regard to anything external, remains always similar and immovable. Relative space is some movable dimension or measure of the absolute spaces; which our senses determine by its position to bodies; and which is vulgarly taken for immovable space; such is the dimension of a subterraneaneous, an æreal, or celestial space, determined by its position in respect to the earth.

This text is important for our story, notably for the insistence that Newton places on the difference between *absolute* time and space and *relative* time and space, which are in relation with the external, palpable world: that world which the human senses can perceive. On the other hand, the revolution ushered in by Einstein will differ from this by, among other things, thinking of time and space in relation to what human senses, or measuring instruments like clocks and rulers, can perceive. This is one of the origins of the name *theory of relativity*, given to the point of view introduced by Einstein in June 1905 (and generalized in 1915, as we shall see).

Newton's insistence on putting the concepts of space and time on a pedestal had three motivations. First, from other texts of his we see that he essentially identified absolute time with God's duration or time such as God experiences it, and absolute space with the sensorium of God, or, in essence, the body of God, since sensorium refers to the medium through which one receives sensations. Second, he hoped to differentiate his ideas from those of his eternal rival, the great Continental scientist and philosopher Gottfried

Wilhelm Leibniz. For Leibniz, time and space are not real, absolute entities, but pure relations between substances. Let's read what Leibniz wrote to Newton (through Newton's substitute: Clarke):[10]

> As for me, I have remarked more than once, that I hold SPACE as being something purely relative, like TIME; it is an order of co-existences, just as time is an order of successions.

Finally, and above all, Newton, who was a physicist of genius, understood that by postulating, *a priori*, the "absolute and mathematical" concepts of space and time, he would give to the physics of his era tools of a power that had never before been seen. This power is testified to by the extraordinary advances initiated in his *Mathematical Principles of Natural Philosophy* (and developed in the three centuries that would follow). Einstein indeed realized explicitly that Newton had forged his way forward through the only route conceivable in his era. As Einstein wrote near the end of his life, in a passage[11] showing his instinctive intimacy with Newton, across the centuries: "Newton, forgive me; you found the only way which, in your age, was just about possible for a man with the highest powers of thought and creativity."

The vision of "relative" space and time proposed by Leibniz is quite profound, and remains relevant today (as we shall see). In his letters to Clarke-Newton, Leibniz gave several arguments in favor of his proposal. His objections to the Newtonian concepts were founded on certain properties of in-observability of absolute time, absolute space, or absolute motion. One of these properties of in-observability would, in fact, be one of the basic postulates of Einstein in 1905: the *principle of relativity*.

The essential idea of this principle can already be found in the work of Galileo (who nevertheless did not use this terminology, which would only be introduced later). He gave a very intuitive and explicit formulation of this idea in his *Dialogue Concerning the Two Chief World Systems*. There, he asks his readers to shut themselves up in the cabin of a ship and, remaining in this closed space with no view to the exterior, to observe the behavior of various processes. He suggests, for example, to observe the way in which flies and butterflies flutter, or the way a bottle might empty itself drop by drop into a vase placed underneath it. When the ship is at rest, the butterflies flutter every which way, with no preference for any particular direction, and the drops fall vertically from the bottle. Galileo then asks his readers to make the same observations when the ship is moving, with a speed as fast as one may wish, provided that the motion is uniform, with a constant speed and direction, without any rough water or changes in tack. He asserts that any processes observed within the interior of the cabin will occur in precisely the same manner as when the ship was at rest. Whatever the speed of the ship, the butterflies shall still flutter every which way, with no preference in direction, and the drops shall always fall exactly vertically. In other words, the uniform, collective motion of the ship is *undetectable* for an observer who participates in this motion, and who only observes what happens inside the cabin. Today's reader may adapt Galileo's description to a more modern method of transport. For example, one may consider an airplane, with the window shades down, where (in the absence of turbulence!) it is impossible to detect, through observing the behavior of processes inside the cabin, the motion of the airplane. Galileo further summed up the principle of relativity in a very concise way: "the motion is as nothing." Here it is understood that one is speaking of a uniform, collective motion, at a constant speed and in a straight line.

Keeping this in mind, and knowing as well that this undetectability of uniform collective motion is a consequence of Newton's axioms on motion, let's now listen to what Leibniz says:

> One now replies that the reality of motion is independent of the observation, and that a ship can move forward without those inside ever noticing. I respond that the motion is independent of the observation, but that it is not independent of observability. There is no motion when there is no observable change.

This insistence of Leibniz (which he based on his *principle of the identity of indiscernibles*) on the need for a criterion of *observability* of all supposed reality is quite modern, and might have inspired the point of view introduced by Einstein in 1905. However, Leibniz's objections did not impress the Newtonians. It must indeed be recognized that, faced with the monumental Newtonian synthesis, which succeeded in explaining the main features of solar system mechanics, and which allowed for the mathematical formalization of an ever-growing number of physical phenomena, Leibniz's approach, although conceptually seductive, had little to offer in terms of the practical description of physical phenomena. Leibniz's critique of the concepts of absolute time, space, and motion did, however, continue to make its way in an underground manner through subsequent thought. In particular, it resurfaced (in a new form) in the writings of the Austrian physicist and epistemologist Ernst Mach, which Einstein read and discussed with great interest in Bern, with the members of his discussion club: the Olympia Academy.[12]

Thus, from the beginning of the eighteenth century, physical reality was thought of within the framework introduced by Newton. Space and time are absolute. They are believed to exist independently of matter. They define reality's *container*, while matter comprises the *contents*. If one were to get rid of all the matter, there would still remain a large, empty frame: space, like an empty stage after the actors have left. And on this empty stage, time would continue to flow uniformly, despite the absence of any human consciousness, or any clock, capable of experiencing or measuring it. This conception of absolute time seemed natural, because it seemed to be a simple generalization of the psychological time experienced by every individual. The concept of absolute space, on the other hand, seemed psychologically disturbing, for the usual concept of space, formed by individuals through what they observe, is constructed from our experience as beings attached to the Earth. Everyday space is necessarily rooted in a ground, from whence comes a certain unease caused by the groundlessness that Newtonian absolute space represents. This feeling of groundlessness had been one of the causes of the resistance to the Copernican ideas concerning the motion of the Earth around the Sun. The Copernican revolution, continued and embellished by the ideas of Galileo, Kepler, Descartes, and Newton, had thus led the way to a conception of space that was *rootless* and *ghostly*. The unease caused by the notion of absolute space (contemplated in all its nudity, in the absence of matter) can be seen in this beautiful passage by James Clerk Maxwell:

> There are no landmarks in space; [...] We are, as it were, on an unruffled sea, without stars, compass, sounding, wind or tide, and we cannot tell in what direction we are going.

Ether: The Materialization of Absolute Space

Fittingly, it was Maxwell himself—he who had worried about the groundless-ness caused by the in-observability of space—who made an essential contribu-tion to the transformation from Newton's abstract, absolute space to a phys-ical, quasi-concrete medium: the ether. In the beginning of the nineteenth century, the brilliant French physicist Augustin Fresnel (who died, alas, at age 39) showed, by a combination of experimental and theoretical results (some of which had been prefigured by the Englishman Thomas Young), that light was a wave-like phenomenon. In contrast to Newton, who had developed a theory in which light was composed of particle-like *corpuscles* emitted by luminous bodies, Fresnel convinced his contemporaries that light was, in fact, a *wave*.

In the same way that an insect which disturbs the surface of a tranquil lake will create waves which ripple out in every direction, moving along the surface of the water, Fresnel had the idea that luminous bodies disturbed an elastic medium filling all of space, thus starting waves which then moved at a speed of 300,000 kilometers per second. This medium of light-wave propagation was called the *luminiferous ether*, meaning light-carrying ether. Forty years after Fresnel's work, Maxwell, in the course of his fundamental research into the combined evolution of electric and magnetic fields, discovered something quite remarkable: electric and magnetic fields moved through empty space in the form of an *electromagnetic wave* in which the electric and magnetic fields played (as described earlier) leap-frog with each other. The structure of these electromagnetic waves was of exactly the same type that Fresnel had been forced to postulate in order to explain the behavior of light.[13] Even more remarkably, by using certain electric and magnetic measurements, Maxwell was able to calculate and predict the speed of propagation of electromagnetic waves: he found (approximately) 300,000 km/s. This convinced him that light was nothing but a particular variety[14] of electromagnetic waves,[15] and led him to suggest the idea that the same elastic medium served for the propagation of light waves and electromagnetic waves. The ether thus acquired a central importance in physics. It became an invisible material substance, transparent and penetrable (although quite rigid) by all ordinary bodies. The ether filled all of space, and stayed at rest. It served not only as a propagating medium for light and electromagnetic waves, but also as a place of existence for all the forces acting on ordinary matter: the gravitational, electric, and magnetic forces. For example, the "lines of magnetic force" revealed by iron filings around a magnet were thought to be caused by a particular substructure within the ether. At the end of the nineteenth century, a good number of physicists even believed that ordinary matter was nothing but highly concentrated ether, and thus that "everything was ether."

For our story, we note that, above all, the ether had (in the end) furnished a "ground" which could replace the immobile Earth of the pre-Copernican world. By filling and materializing Newton's absolute space, it put an end, in principle, to the unease caused by the in-observability of this abstract space, that unease evoked in Maxwell's earlier quote: "[...] an unruffled sea, without stars, compass, sounding, wind or tide ...." For all of these reasons, both scientific and psychological, the community of physicists at the end of the nineteenth century was absolutely convinced of the existence and reality of the ether. It is amusing in this regard to read the definition of the word "ether" in a French dictionary of the time, the *Nouveau Larousse Illustré* (published around 1903, just before the Einsteinian revolution):

> Ether [...] *ether* is a term for an invisible, impalpable and imponderable element, spread everywhere, in empty space as well as in the interior of both transparent and opaque bodies, and whose existence, for a long time hypothetical, now seems to enjoy all the characteristics of scientific certitude ...

It is ironic that this definition, insisting on the "scientific certitude" of the existence of the ether had been written just before a young "technical expert, third class" of the Bern Patent Office would usher in modern physics by affirming (among other things) the absence of the ether! At any rate, this passage shows how prevalent the belief in ether was at the beginning of the twentieth century, and helps one to appreciate the intellectual audacity of the young Einstein, who was ready to overturn some of the most well-established scientific dogmas of his time.

Butterflies in a Ship's Cabin

To conclude our recollections concerning the conceptual framework considered as obvious by all of Einstein's contemporaries, we shall now discuss the relationship which they believed to hold between the behavior of physical processes as seen from a moving reference frame, and the behavior of these same processes as seen from a reference frame at rest. To be concrete, let us return to the situation imagined by Galileo: a ship's cabin, within which a whole range of processes occur, like the fluttering of butterflies. The shore here plays the role of a reference frame at rest, as opposed to the ship (and its cabin) which constitutes a moving reference frame. We are thus interested in the link between the description of the "proper" motion of a butterfly, as seen by an observer in the interior of the cabin, and the description of the motion of the same butterfly, as seen by an observer on the shore. In particular, we ask ourselves the following question: if the ship moves along the shore at a speed of (say) one meter per second, and if a butterfly moves with respect to the cabin (and towards the front of the ship, parallel to the shore) with a speed of two meters per second, at what speed does the butterfly move with respect to the shore? The scientists of the beginning of the twentieth century would have responded to this question in the following way.

Given Newton's notion of absolute time, which coincides with our intuitive notion of a unique time for all beings and all physical processes, the observer on the shore and the observer in the cabin (as well as the butter-fly!) experience the same flow of time, or duration. Thus, if one considers what happens "during one second," which here means one second of universal time, seen at the same time by the observer on the shore and the observer in the cabin, it is "obvious" that "during this second" the boat shall have advanced by one meter with respect to the shore, and the butterfly shall have advanced, in the same direction, by two meters with respect to the cabin. However, given Newton's absolute notion of space, it is also "obvious" that one meter measured in the cabin is equal to one meter measured at the shore. Thus, "during one second," the butterfly will have advanced with respect to the shore by the *sum* of one meter and two meters, that is $(1 + 2 = 3)$ by three meters. The speed of the butterfly with respect to the shore is thus three meters per second. The resulting speed is thus simply given (in general) by the *sum* of the "proper" speed with respect to the cabin and the "drifting" speed of the ship (and of the cabin) with respect to the shore. This law of addition of speeds applies very generally to any moving body, as well as any propagating phenomenon. For example, if one replaces the butterfly by a ray of light, one then deduces that the speed of a light ray with respect to the shore is the sum of the speed of the ship and of the speed of the ray of light with respect to the cabin. Thus, as soon as the ship has a nonzero speed, light cannot have the same speed both with respect to the shore *and* with respect to the ship.

The scientists of the early twentieth century all[16] saw the ether as a sort of universal "shore," or "unruffled sea, without ... wind or tide," defining a reference frame truly at rest. On the other hand, they thought of the Earth as a sort of ship in perpetual motion on this sea, continually changing its speed as it orbits around the Sun. Since, by definition, light propagated through the ether with a well-defined speed of 300,000 kilometers per second, they concluded that the speed of light with respect to the Earth was the *difference* between 300,000 kilometers per second and the "absolute" speed of the earth with respect to the ether. In particular, the speed of light measured on Earth should be slightly different from 300,000 kilometers per second, and should vary with the seasons, as well as with the relative orientation between the direction of the "absolute" motion of the Earth and the direction of the light ray considered.

This prediction (which was made in particular by Maxwell himself) finally offered the possibility of obtaining evidence for the reality of the ether. Such a result would put an end to the unease associated with the inobservability of Newtonian absolute space, which had tarnished its conceptual beauty, as highlighted by Leibniz's subtle reasoning. Great effort was therefore deployed, at the end of the nineteenth century, to detect the motion of the Earth with respect to the ether. In particular, one of the most precise experiments performed was that of the American physicist Albert Michelson, which he completed at Potsdam, just outside of Berlin, in 1881. He repeated this experiment with greater precision in the United States in 1887, in collaboration with the chemist Edward Morley. Although the precision of Michelson's measurements were quite sufficient to detect the expected alteration of the speed of light due to its combination with the motion of the Earth, Michelson had the great surprise of detecting nothing at all! This completely unexpected result forced the leading scientists of the time, including the Irishman George FitzGerald, the Dutchman Hendrik Lorentz, and the Frenchman Henri Poincaré, to invent various hypotheses to explain the nonobservability of the motion of the Earth with respect to the ether. This led them to publish some scientific papers which anticipated some of the content of Einstein's work of June 1905.

"The Step"

At the end of his life, in conversation with the man who would become his most penetrating biographer, Abraham Pais, Einstein referred in an impersonal fashion to the (special) theory of relativity, calling it *den Schritt*, that is, "the step." Note that Einstein does not speak of a "jump."[17] Nevertheless, he knew that in June 1905 he had introduced a major conceptual break with previous thought. But he also knew that this conceptual break was subtle, and essentially programmatic. In other words, it paved the way for future work, rather than immediately bringing a harvest of new, concrete results. In fact, even if Einstein was unaware of it for the moment, the greater part of the mathematical formulas contained in his June 1905 article had already been written not long before, or even concurrently, by Lorentz and Poincaré. Moreover, some of the physical ideas contained in Einstein's article can be found, separately, in the work of some of his contemporaries.[18] This fact has periodically led certain authors[19] to revisit Einstein's contributions, and to minimize them by comparing them with extracts from the work of his predecessors. This approach is founded on a fundamental misapprehension of the conceptual step made by Einstein in June 1905. Keeping with the spirit of this book, we shall try to make an appreciation of the essential novelty of the point of view adopted by Einstein, without entering into technical or historical details.

The essential point is that Einstein rejects the Newtonian concepts of absolute time and absolute space, as well as the concept of ether, which was traditionally identified with absolute space. It required great audacity to reject in its entirety a conceptual framework which had been so useful to the development of physics for centuries and which, above all, had become so familiar, and seemed to correspond so well to the intuitive idea of space and time shared throughout the world.

And how did he go about rejecting these Newtonian concepts? Did he reject them immediately? With what did he replace them? By a space and time that were purely relative, in the sense that Leibniz wanted? No. In fact, Einstein proceeded in a very gradual fashion. He begins by assuming the existence of a reference frame at rest, with respect to which the usual concepts of space, in the sense of ordinary Euclidean geometry, and of time, in the Newtonian sense, are valid. He assumes that in this reference frame at rest, Newton's laws of dynamics are valid in the first approximation, and that light in this reference frame propagates[20] at a speed of 300,000 kilometers per second. This point of departure was completely conventional, and acceptable by every physicist of his epoch (it sufficed to identify Einstein's "reference frame at rest" with the ether). Einstein here contents himself with simply clarifying that in order to give a clear observational sense to the notion of time within this reference frame, one must first (mentally) provide each point in this reference frame with a clock.[21] For, he remarks, a clock can only determine the time of events that occur where the clock itself is situated. He continues his remarks by noting that one must "synchronize" all of the clocks with each other, in the same way that two people synchronize their watches by adjusting the needles of one of the watches until they show the same time. He proposes to accomplish this synchronization between distant clocks by exchanging light signals and then compensating for the time taken by the light in traveling between them, calculated by dividing the distance between the two clocks by the speed of light, 300,000 kilometers per second. Up until now, nothing is new. Every physicist would have agreed with this fashion of proceeding, which was at any rate quite similar to the method of clock synchronization by telegraphic signals, used in a routine fashion at the end of the nineteenth century.[22]

It is at this point that the great novelty of Einstein's approach appears, when he introduces the strict validity, for all laws of physics, of the principle of relativity. His understanding of this principle is exactly the same as in Galileo's text cited above: undetectability of any difference whatsoever between the behavior of local physical processes observed in a reference frame at rest, or in a reference frame in uniform motion with respect to the reference frame at rest. Among the laws of physics for a reference frame at rest, he has already stated that light propagates at a fixed speed (300,000 kilometers per second), independently of the direction of propagation and of the speed of the source. Thus, the principle of relativity implies, as a first consequence, that light must also propagate at the fixed speed of 300,000 kilometers per second in any reference frame in uniform motion with respect to the reference frame at rest.

Having arrived at this stage of Einstein's reasoning, every physicist of his epoch must have begun to say to himself: "That's absurd! The well-known law of the addition of speeds implies that the speed of light in a reference frame in motion would be the *difference* between its speed in the reference frame at rest and the overall speed of the moving reference frame. This difference could never be again equal to 300,000 kilometers per second." However, Einstein continues by explaining that, essentially,[23] the well-known law of addition of speeds is, in fact, the consequence of *tacit hypotheses* on the behavior of rulers and clocks in motion that have no *a priori* reason to be true. Recall the example above of the butterfly in the cabin of a ship. If one strips away, layer by layer, this example, where the boat advanced at "one meter per second," "while" the butterfly advanced at "two meters per second," we see that it was actually necessary to make three separate hypotheses to calculate the speed of the butterfly with respect to the shore. First one had to suppose that a distance of two meters traveled by the butterfly in the cabin was equivalent to a distance of two meters seen from the shore. Secondly, one had to suppose that a duration of one second passed within the cabin was equivalent to a duration of one second passed on the shore. The third hidden assumption was that cabin and shore could be considered "at the same instant." In other words it was assumed that the notion of *simultaneity* was the same both in the cabin and on the shore. However, Einstein says that there is no *a priori* reason that a ruler measuring one meter when it is viewed at rest should still appear to have a length of one meter when it is viewed in motion. Likewise, a clock ticking every second when it is viewed at rest has no *a priori* reason to appear as ticking each second when it is seen in motion. Finally, Einstein demonstrates through simple reasoning that if one defines simultaneity in all reference frames through the process of synchronization of separated clocks by the exchange of light signals (assumed to travel at the fixed speed of 300,000 kilometers per second), then two events which occur "at the same instant" in a reference frame in motion *are not simultaneous* with respect to the reference frame at rest (and vice versa).

Having thus rattled centuries-old convictions about space and time, Einstein then shows how to construct new notions of space and time. To do this, he uses a very powerful tool: the principle of relativity itself. In imposing the exact validity of this principle, he deduces how a one-meter ruler in motion must appear to an observer at rest, at what frequency a clock in motion seems to beat when it is compared to a clock at rest, and how the notion of simultaneity changes between two reference frames in relative motion. He thus arrives at a collection of mathematical formulas connecting the coordinates of space and time (length, width, height, and date) for a particular event, according to whether it is seen from a reference frame at rest, or from a reference frame in uniform motion.[24] Einstein was probably unaware that these mathematical formulas had been written previously, notably by Lorentz, and that they had been studied in detail by Poincaré.[25] This is hardly important. The essential point is that the physical meaning of these equations was completely new in Einstein's work. Indeed, not one of Einstein's contemporaries questioned Newton's absolute space and time. For them, among the diverse variables which entered into these equations, only the lengths, widths, heights, and dates in the reference frame at rest, defined by absolute space (and the ether), represented *real* coordinates of space and time. The other variables were either *apparent* coordinates, or simple mathematical intermediaries. We shall return later to this crucial difference between Einstein and his contemporaries.

After having obtained the equations which connect the space and time coordinates of a single event as observed from two different reference frames, Einstein discusses their physical interpretation. He then derives the modified form taken by the law of addition of speeds in his new theory. He finally verifies that this new law is really such that in "adding" an arbitrary subluminal speed to the speed of light, one obtains, once again, a speed equal (in magnitude) to the speed of light. The circle is now complete: he has shown the compatibility between the principle of relativity and the principle that light always propagates at the same speed. Moreover, all of this theoretical development (as well as the rest of the article) renders the concept of luminous ether, as well as of absolute space, completely "superfluous." Einstein does not say any more on this, but this simple remark amounts to a death sentence for all those old concepts considered obvious by his contemporaries.

This concludes our commentary on the first part of Einstein's article of June 1905. He described this first part as being "kinematical," concerning the properties of space, time, and motion. In the second part, called the electro-dynamical part, he shows how the application of his new approach concerning the properties of space and time to Maxwell's equations of electromagnetism permits one to reduce any electromagnetic or optical problem treating bodies in motion to a series of problems with bodies at rest. He also obtains, in passing, a number of new and important results. He finishes by showing how the principle of relativity obliges one to modify Newton's second law of dynamics (connecting the force acting on a body, its mass, and its acceleration). He deduces in particular that the usual expression for the kinetic energy of a moving body, in other words the energy associated with the motion of a body, must be modified when the body moves at a high speed. He finds that the kinetic energy grows without bound when the speed of the body approaches the speed of light. This permits him to conclude that the speed of light is a limiting, unattainable speed. Finally, Einstein had found the answer to the question which had eluded him since he was 16 years old! It is impossible for an observer to catch up with a ray of light. The speed of the observer is necessarily smaller than the speed of light. Moreover, even if his speed, in a particular reference frame, is extremely close to that of light, the observer shall still see, in his own reference frame, that light continues to move away from him at 300,000 kilometers per second. Run as fast as you wish—you can never catch up with light.

Time Deregulated

Let us return to the essential conceptual innovation of the theory of relativity, born fully formed from Einstein's brain in the spring of 1905. As Einstein told Besso, when he saw him again the day after their crucial discussion: "Thank you. I have completely solved my problem. An analysis of the concept of time was the solution. Time cannot be absolutely defined, and there is an inseparable relation between time and signal velocity."

This new understanding of the concept of time which Einstein introduced is also what distinguishes Einstein's work from the contributions of all the other scientists (and in particular Lorentz and Poincaré) who were then working on the electrodynamics of moving bodies. For Lorentz and Poincaré, there was only one true time, the absolute Newtonian time with which they had always been familiar. The other variables, which looked like time but were associated to reference frames in motion, remained simple mathematically convenient auxiliaries. This is confirmed by what Einstein wrote in 1907:

> It turned out, surprisingly, that it was only necessary to define the time concept precisely enough to overcome the ... difficulty. All it needed was the realization that an auxiliary term introduced by H. A. Lorentz and called by him "local time" could be defined as "time," purely and simply.

This difference is also highlighted by what Lorentz himself wrote in 1915:

> The chief cause of my failure [in discovering relativity] was my clinging to the idea that only the variable t can be considered as the true time, and that my local time t' must be regarded as no more than an auxiliary mathematical quantity.

In Poincaré's case, things are more subtle, since Poincaré was the first to understand, in 1900, that Lorentz's "local time" t' was more than simply a useful auxiliary quantity. Poincaré had indeed realized that if observers in motion decided to synchronize their watches by exchanging light signals, with the assumption that the duration of the signal transmission between the two observers is the same in both directions, their watches would show, at least to the first order of approximation, Lorentz's "local time" t'. In spite of this important understanding, Poincaré discussed this procedure of synchronization in 1904 by saying:[26]

> The watches thus constructed will therefore not show the true time, they will show what might be called local time, with the effect that one of them will run late with respect to the other. It hardly matters, since we have no way to detect it. All of the processes which occur at A for example will run late, but all to an equal extent, and the observer [located at A] will not perceive it since his watch is running late as well. Thus, as required by the principle of relativity, there will be no way to know if one is at rest or in absolute motion.

This quote shows clearly that, for Poincaré, this method of synchronization, as practical (and widely practiced at the time[27]) as it may be, only defined a deformed approximation to "real" time—Newton's universal, absolute time—since it assumes a symmetry in the duration of signal transmissions which he believed not to be true for observers in "absolute motion." Indeed, as Poincaré said in the sentence preceding the quote above:

> In the contrary case, [when the observers A and B are not "fixed,"] the duration of transmission will not be the same in both directions, since station A,

for example, moves towards the optical perturbation emanated by B, while station B moves away from the perturbation emanating from A.

The words used by Poincaré, "true time," "run late," "absolute motion," "fixed," etc., show explicitly that his thinking remained safely inside the horizon of the Newtonian concepts: absolute time, absolute space, and absolute motion.

A crucial consequence of the limitations of Poincaré's conceptual framework is that the "local time" of which he speaks in the 1904 text cited above differs in an essential way from the "time" that Einstein attributes to a moving reference frame. Indeed, a careful reading of Poincaré's 1904 text, of the course[28] that he gave at the Paris Faculty of Sciences during the winter of 1906–1907, and of an article[29] published in 1908, show that the "time," let's call it τ, of which Poincaré speaks is always a time whose second is measured by clocks in "absolute rest." Because of this, although Poincaré anticipates Einstein in speaking of synchronization through the exchange of light signals, the "Poincaré time" τ is *larger* than the "Einstein time," let's call it t', by a factor k which depends on the "absolute" speed of the moving observers A and B.[30]

"Who cares about this numerical factor k?" the reader of this book is no doubt thinking! Even more so since in Poincaré's later mathematical work, instead of using the time τ which he defined in the above passages, he used the time $t' = \tau/k$, as defined by Einstein—and Lorentz before him. Nevertheless, this numerical factor k is absolutely essential, since it reflects the gigantic conceptual gulf between the thought of Poincaré (and of Lorentz) and that of Einstein. Indeed, Einstein deduced, in his article of June 1905, a remarkable observable consequence deriving from the presence of this factor: a moving clock will not tick with the same rhythm as a clock at rest. More precisely, the rhythm of a clock moving at a speed v with respect to a certain reference frame, seems—when observed from this frame, and after removing the delay due to the transmission of electromagnetic signals—to be *slower*, by a factor $k = 1/\sqrt{(1 - v^2/c^2)}$, than the rhythm of a clock of identical construction at rest in this reference frame. For example, if the speed v is 86.6% of the speed of light, around 260,000 kilometers per second, the moving clock will appear to have a rhythm *two* times slower than a clock at rest (in other words, such a speed corresponds to a factor of k equal to 2). If a clock, seen at rest, ticks once every second, the same clock will seem to tick once every two seconds, when one observes it moving at this high speed. This new physical effect, generally called *time dilation*, was never imagined to exist before Einstein. While some of the equations manipulated by Lorentz and Poincaré were identical to those derived (independently) by Einstein, and indeed contained this factor k modifying the second measured by clocks in motion, Lorentz and Poincaré always thought of time in terms of Newton's absolute time. They never suggested, as Einstein did, that a moving clock would tick at a different rate than that of a clock at rest.[31]

We have here reached the very heart of the conceptual innovation made by Einstein's theory of relativity: the dethroning of a Newtonian absolute time shared by the entire universe, and its replacement by a multiplicity of individual times, each disagreeing with the other. This great deregulation of time is strikingly illustrated by what is called the *twin paradox*. A first version of this paradox was suggested by Einstein at a conference in Zurich in January 1911. He imagined enclosing a living organism within a box, and then giving this box a speed very close to that of light. (We note that the factor k relating the time on Earth and the time inside the moving box approaches infinity when the speed of motion approaches the speed of light.) Once the box has traveled a great distance, let's say five light-years, the box makes a return voyage back to its departure point, still moving at a speed very close to that of light. Once returned, after opening the box, one finds that the traveling organism will have barely aged, while similar organisms, which have remained immobile on Earth, will have aged ten years (or, in the case of a much longer journey or of an organism with a fairly short life-span, will have long ago given way to new generations).

The French physicist Paul Langevin rendered Einstein's parable more picturesque by imagining that the voyaging organism is a man, shot out in a cannonball à la Jules Verne, who upon his return finds, like Rip Van Winkle, that his contemporaries have aged considerably, while he has hardly aged at all. In modern versions of this paradox, the cannonball is usually replaced by a rocket, and one imagines a pair of twins, one of which voyages and then returns to find that his sedentary brother is now much older than he is. Note in particular that when we speak of aging, or of time passing, we are speaking of "time, pure and simple," as an organism experiences it, counted, for example, in the number of heartbeats or the number of breaths taken.

Temporal Refrigerators

Whichever version of the twin paradox one chooses, the effect of the deregulation of time described by the factor k is only detectable if the voyager moves at a speed comparable to the speed of light, 300,000 kilometers per second. This is a gigantic speed, compared to the speeds to which we are accustomed. Because of this, the effect described by the twin paradox is only relevant in situations too far from our intuition to be able to affect our age-old understanding of time. We can nevertheless amplify the psychological and existential impact of this paradox by following the example made by the Russian-born physicist George Gamow in his excellent popularization novels.[32] We start by imagining that we live in an alternate universe, similar to ours in every respect, except that the speed of light is very much lower. For example, we imagine that the speed of light is only 30 kilometers per hour. In such a universe, the outer part of a merry-go-round may turn with a speed very close to the speed of light. Such a carousel would then form a sort of *temporal refrigerator* that would freeze the flow of time for people on the merry-go-round, compared to the flow of time experienced by stationary observers. For example, if a mother of two twins made only one of her children get up on one of the wooden horses, and then forgot him there for one year(!), she would find him, at the end, practically unchanged in age, while the child who stayed on the ground will have aged one year (along with his mother). We note that such a *temporal refrigerator* would not allow one to "live longer," in the sense of being able to live through a greater number of heartbeats than the number one would experience without getting on the carousel. The total time experienced by the traveling twin, in terms of the number of heartbeats, will be the same (neglecting the biological effect of the centrifugal force of the carousel) as that seen by a sedentary twin. Mounting and remaining on the carousel would only permit us, similarly to cryogenic freezing, to return to the world much later and find that others have lived a number of years that we have not lived.

The twin paradox illustrates clearly the fact that Einstein's theory of relativity completely overturns the common notion of an impersonal time, through which the entire evolution of the world flows. This conceptual upheaval had, in April 1911, struck philosophers from around the world when Langevin delivered a lecture on the evolution of space and time at the Bologna International Congress of Philosophy, where he presented his parable of the voyager sent at great speed inside a cannonball. In the '20s, the general public the whole world over was deeply intrigued when the newspapers began to mediatize Einstein and his revolutionary theories of space and time. It is quite astonishing that today, a century after Einstein's original work, this conceptual upheaval is no longer generally understood as such. The greater part

Figure 1. The twin paradox.

of current journalistic treatments of the consequences of Einstein's theories, such as the big bang or black holes, in fact, presuppose the common notion of universal time. Far from being assimilated by the general public, Einstein's conceptual revolutions are simply ignored. One of the motivations for this book is to try to return to them some part of their original vigor.

We shall continue in the next chapter our discussion of the deep significance of Einstein's deregulation of time. Let us finish this chapter by simply indicating that the observable phenomena associated with the deregulation of time, predicted by Einstein and connected to the presence of the velocity-dependent factor k, were indeed verified some years later. In his original article of June 1905, Einstein indicated that in principle the effect of the deregulation of time produced by moving at high speed should be observable on Earth, by comparing a (spring-driven) clock located at the equator with an identically constructed clock located at the pole. Indeed, if one neglects the effect of gravitation,[33] the rotating Earth is like a carousel, and a clock situated at the outer edge of this carousel must tick at a slower rhythm than a clock positioned at the hub of the carousel. In 1907, Einstein noted that another consequence of this deregulation of time, which he had, in fact, already deduced in his article of June 1905, would be easier to observe: the frequency emitted by an atom moving perpendicularly to the line of sight of an observer is lower than the frequency of a twin atom at rest. This consequence was verified with great precision by Herbert E. Ives and G. R. Stilwell at the end of the '30s. The ability of motion at high velocity to produce a temporal refrigerator was verified in the '40s and '50s by studying the apparent lengthening of the lifetime of certain elementary particles. In the '70s, the stability of atomic clocks became so good that it was possible to directly verify the apparent slowing of clocks traveling in airplanes. Today the effect of the deregulation of clocks in motion is verified and constantly used in the American Global Positioning System (GPS), which uses a constellation of atomic clocks moving in orbit around the Earth. In these two latter cases, the effect of deregulation due to relative movement is combined with an effect due to gravitation which we shall speak of later. In fact, the stability of present-day commercial atomic clocks is so great that it would probably be possible to perform a version of the twin paradox in front of the general public, by comparing a clock on a laboratory carousel to its twin on the ground. The reader may have perhaps concluded from the twin paradox that jogging regularly would, thanks to the slowing down of time for a body in motion, permit one to stay younger, or rather to grow old less quickly, than one's sedentary friends. This may be true but, unfortunately, the speed of light is so great with respect to our running speed, that the gain in time is minuscule. For example, someone running night and day for 75 years at the speed of a marathon racer would only gain one-third of a microsecond!

To conclude, let us comment on the use of the word "paradox" to characterize the effect of the relative deregulation of two clocks in motion with respect to each other. In Greek, *paradoxos* signifies "contrary to common opinion." For a long time, the twin paradox has effectively shocked our preconceptions by clashing with common sense.[34] Many have doubted its reality. Today, we know that this effect is real, and it would be more suitable to modify common sense to adapt to a new concept of time. When that is finally done, one could perhaps replace the expression "twin paradox" with "twin parable" or "twin paradigm."

An Infinite Multiplicity of Desynchronized Times

To sum up, Einstein's work of June 1905 overturned the concept of time that had been held for centuries. As Max Planck quickly understood, the upheaval produced by the theory of relativity in the physical description of reality is comparable, "in extent and depth," to the Copernican revolution. Planck even went so far as to say that Einstein's "step" was more audacious than anything previously accomplished in speculative science and that, for example, discovering the possibility of non-Euclidean geometries was nothing but "child's play" by comparison.

To make a biological comparison, one could say that before Einstein, time was understood as a unique cosmic pulse, co-extensive with collective reality, and echoing the universal evolution by beating, simultaneously in all of space, every second in a perfectly regular fashion. After the theory of relativity appeared in June 1905, this unique, regular pulse was replaced by an infinite multiplicity of individual pulses, who are not only unsynchronized between themselves, but which in general beat seconds which all differ. How is this gigantic cacophony of discordant pulsations organized? How should one think about the deregulation of all these times? What becomes of the centuries-old notion of the passage of time, traditionally compared to the flow of a river? Such are the questions which find their simplest responses in the notion of a four-dimensional space-time.

L'ILLUSTRATION

RENÉ BASCHET, directeur.   SAMEDI 1er AVRIL 1922   Marcin NORMAND, rédacteur en chef.

LE GRAND PHYSICIEN EINSTEIN A PARIS

Arrivé à Paris le mardi soir 28 mars, le professeur Einstein avait voyagé, depuis la frontière belge, avec M. Langevin, professeur au Collège de France, et M. Charles Nordmann, astronome à l'Observatoire de Paris... « Comment ! écrivait ce dernier dans le Matin du lendemain, c'est cet homme au visage étonnamment jeune, qui a l'air, lorsqu'il rit, d'un étudiant, c'est là celui qui a bouleversé tout l'édifice de la science classique ! »

Einstein in Paris. *L'Illustration*, Saturday, April 1, 1922.

Einstein speaks at the Collège de France, March 31, 1922 (Langevin, seated behind him, whispers any French words he may require). Sketch by Lucien Jonas from *L'Illustration* of April 8, 1922.

2

The World's Checkerboard

Time is a child playing at checkers.
—Heraclitus

"Time Does Not Exist!"

Paris, France, March–April 1922

4:45 P.M., Friday, March 31, 1922. The Latin Quarter. A crowd is pressed against the gates of the Collège de France. Entrance is restricted: only those with an invitation card or a journalist's badge are allowed in. Many curious citizens have nevertheless come to feel the atmosphere of such an exceptional event and with the hope of seeing, from a distance, the world-famous celebrity who has just arrived in Paris. For many days, a serial of a new genre occupies the headlines of the Parisian dailies:[1] "A scientist of genius—Professor Einstein to come to Paris," "A Scientific Event—Einstein in Paris," "Will Einstein Come to Paris?", "Einstein Doesn't Believe in Time and Space but He Believes in Democracy," "Einstein and Relativity—A New Era for Science," "Time Does Not Exist, Says Einstein. But Daylight Savings Time Exists, Said Mr. Honorat. And Tonight, It Springs Forward," "Einstein and the Relativity of the Age," "Understanding Einstein—What is Time? What is Space?", "Einstein Expected in Paris This Afternoon," "Einstein Arrived at Midnight at the Gare du Nord," "The German Physicist Einstein Arrived Last Night in Paris."

5:00 P.M., March 31, 1922. Hall 8 of the Collège de France is full. The room is filled well beyond its 350-seat capacity. Many of the invitees must remain at the door. Rarely has such a density of spectators been squeezed into the Collège de France. As had been the case for the celebrated lectures of the philosopher Henri Bergson, a large part of the general public has tried to obtain entry cards. However, Paul Langevin, a professor at the Collège and Einstein's host, has remained inflexible. He has reserved most of the invitation cards for scientists or for students. Of course, the French scientific and cultural elite are all there: notably the physicists P. Langevin, J. Becquerel, L. Brillouin, and J. Perrin; the mathematicians E. Borel, E. Cartan, J. Hadamard, P. Lévy, and P. Painlevé; and the philosophers H. Bergson, L. Brunschvicg, E. Le Roy, and E. Meyerson. Among the young students invited to attend this exceptional event, we take special note of Alfred Kastler, who was then 20 years old and a student at the École Normale Supérieure (he would later receive the 1966 Nobel Prize in physics for a discovery tied to concepts introduced by Einstein in 1916). Few women are present. Many ladies from high society who have expressed their desire to attend this plenary conference have been refused. Nevertheless, apart of course from Marie Curie, an exceptional scientist and personal friend of Einstein (as well as an intimate friend of Langevin), we notice within the room: the princess Edmond de Polignac, born Winnaretta Singer,[2] who played an important role in French cultural and scientific life through her salon and her patronage, the countess Henri Greffuhle, another important patron, and the countess Anna de Noailles, a famous poet.

5:10 P.M. The atmosphere becomes electric. Finally, ten minutes later than expected, Einstein enters the room, with Paul Langevin and Maurice Croiset, administrator of the Collège de France. To imagine the beginning of this session, as well as to see and hear, almost in real time, Einstein's lecture (originally in French), we pass the floor to the journalist Raymond Lulle, who gave a perspicacious account of the session in the French *L'Œuvre* of April 4, 1922:

> A frenetic ovation, joined into even by those who propose to fight hardest against the hero of the day. A very simple and tactful speech by Mr. Croiset, who shows how the Collège de France has always welcomed the masters of human thought.

> He then gives the floor to Einstein who, clearly quite moved, does not know how to begin. The proximity of Langevin, seated close behind him, seems to give him courage and he enters very simply into his subject.

Before we hear the words spoken by Einstein, we must imagine Langevin, posted just behind Einstein, and ready to whisper any French words which he may require. Einstein begins by noting that even though mathematics serves as an instrument for physics, it does not suffice to simply put physics into equations and then juggle them.

> One must still confront the equations with reality and know what facts the mathematics hides.
>
> "One may very well possess the mathematical apparatus of relativity and understand nothing of the theory itself."
>
> An interesting profession of faith from the mouth of he who many consider as a quintessential abstractor, and one may clearly distinguish the gulf that separates him from some of our mathematicians.
>
> [...]
>
> The language is quite clear, even the simplicity of the vocabulary is evocative. And then, there is the body language: it is that of a sculptor whose hand caresses present, though unreal, forms. His hands are full of forms, moving them, directing them. And he amuses himself prodigiously with his fictional marionettes. His face takes the ebullient air of a child playing tricks.

During Einstein's entire visit, a leitmotiv is going to continually resurface in the newspapers: "Time is No More!", "Time does not exist!", "Illusionary Time," "...Time Is Only a Dream ...." These quotes reflect the interest then taken in the possible impact of science on the philosophical and existential beliefs of the public.[3] And it was not only the newspapers that took seriously the eventual impact of Einstein's theory of relativity on human thought. The whole world over, a number of philosophers reflected on the significance of Einstein's ideas for philosophical concepts. In particular, the philosophers of the "Vienna Circle," notably Moritz Schlick, Rudolf Carnap, Philip Frank, and Karl Popper, as well as the German philosophers Hans Reichenbach and Ernst Cassirer, enthusiastically embraced viewpoints which they motivated through the methods of thought used by Einstein. In France, Léon Brunschvig, Émile Meyerson, and Gaston Bachelard reflected on the philosophical implications of the theory of relativity. Last but not least, there was also Henri Bergson, who was in Hall 8 of the Collège de France, this Friday, March 31, 1922 and who, certainly, felt himself to be quite particularly challenged by Einstein's lecture, which recalled the influence of the theory of relativity on the notion of time.

Bergson and Einstein

The philosopher Henri Bergson had built his entire philosophy on a keen apprehension of the passage of time, experienced in its eternal flow, as an "immediate datum of consciousness." He had deepened this understanding in a series of books: *An Essay on the Immediate Data of Consciousness* (1889), *Matter and Memory* (1896), *Creative Evolution* (1907), and so on. Bergson, and his idealist philosophy founded on the concept of *duration*, reigned as the master of the French philosophical scene. The Parisian intelligentsia hurried to attend his courses at the Collège de France, where he taught from 1900 to 1921. Bergson was therefore particularly challenged by the presentation of the physicist Paul Langevin at the 1911 International Congress of Philosophy in Bologna, concerning the evolution of space and time. In his presentation, Langevin sketched the radical new upheaval of the concepts of space and time caused by Einstein's ideas. In particular, as we have already noted, it was there that he introduced the *twin paradox*, in the form of a traveler inside a cannon-ball, ejected from the Earth and then returning to it, all at an extremely high speed. After his return, the voyager has only lived through two years, while 200 years have passed on Earth.

This *elastic*, mutable behavior of experiential time was in stark contrast to the common notion of a universal time, in step with the evolution of the universe. Bergson, who had founded his entire philosophy on an understanding of time as *duration*, that is, as a flow, or more precisely as a fundamental mobility grasped by the consciousness in its immediacy, must have felt, starting in 1911, that Einstein's time was in opposition to the cardinal concept of his entire philosophical enterprise. The general public was not alerted to the Einsteinian upheaval of time until after 1919, when the media began to take interest in Einstein and his theories. Moreover, as shown by the headlines quoted above from the newspaper articles consecrated to Einstein's visit to Paris, the journalists had deduced from what they understood of his theories that time did not exist, that its apparent flow was nothing but an illusion. It was thus an important challenge for Bergson to interpret the philosophical meaning of Einstein's theories. All the more so since, in the months before Einstein's visit to Paris, Bergson had been hard at work completing a new book entitled *Duration and Simultaneity: Concerning Einstein's Theory*.[4]

Let us quote a passage from the preface to this book, written by Bergson himself:

> We wanted to know to what extent our conception of duration was com-
> patible with Einstein's views on time. Our admiration for this physicist, the
> conviction that he has not only brought us a new physics but also some new
> ways of thinking, the idea that science and philosophy, although different
> disciplines, are made to be complementary—all of this inspired in us the

desire, and indeed imposed on us the duty, to proceed with a confrontation. But our research appeared immediately to offer a more general interest. Our understanding of duration in fact encoded a direct and immediate experience. Without implying as a necessary consequence the hypothesis of a universal Time, it harmonizes with this belief quite naturally. To some extent it was thus everyone's ideas that we wanted to confront with Einstein's theory. And the place where this theory seems to clash into common opinion seems to be found at the most basic level: we shall have to burden ourselves with the "paradoxes" of the theory of relativity, with the multiple Times which flow more or less quickly, with the simultaneities which become successions and the successions simultaneities when we change our point of view. These theses have a well-defined physical meaning: they tell that which Einstein has read, with inspired intuition, in Lorentz's equations. But what is their philosophical significance?

Let us stop to admire this beautiful profession of faith from a sincere and deep philosopher, looking to understand the existential meaning of modern physics. Nevertheless, despite the respect owed to Bergson and his thought, the content of this book is intellectually disappointing, at least for the general public, and even scientifically false, as we shall see below.

For now let's return to Einstein's visit to Paris in the spring of 1922. We have left Bergson in the middle of the crowd which has come to hear Einstein's first lecture at the Collège de France. It is not there that a dialogue could be established between Einstein and Bergson, or any of the other French philosophers who came to listen to him. Such a dialogue was organized one week later, April 6, 1922, at a meeting of the French philosophical society in which Einstein participated. A detailed account of this meeting has been published,[5] and it is an engrossing description. In particular, we can read Bergson's long intervention, in which he tries to summarize, in front of Einstein, the central idea of his book *Duration and Simultaneity*, which had not yet appeared but would soon be in press. (Curiously, Bergson never alludes to the existence of this book.)

This idea is the following: "Common sense believes in a unique time, the same for all beings and for all things [...] Each of us feels themselves to experience duration [...] there is no reason, we think, that our duration is not as well the duration of all things." This idea of a universal time, common to consciousnesses and to things—is it incompatible with the theory of relativity, and its multiple times? Bergson affirms that the answer is no, and he concludes that "the theory of relativity contains nothing incompatible with the ideas of common sense." This conclusion, which put to an end a long, rather obscure presentation of the way in which Bergson interpreted the physical notion of simultaneity, left Einstein fairly speechless. Einstein contented himself with simply commenting that there was no reason to believe that there existed something totally apart from ordinary reality, and which would be a philosopher's time, different from physicist's time. Rather, "the philosopher's time, I believe, is both a psychological time and a physical time; on the other hand physical time could derive from the time of consciousness." Otherwise stated, Einstein politely put in doubt the soundness of the attitude, which he had obscurely perceived in Bergson's somewhat unclear exposé, which consisted of scorning certain scientific advances in the name of an *a priori* philosophical assumption.

In other words, the dialogue between Bergson and Einstein—which might have become more heated if it had led to a detailed confrontation between their points of view, and in particular if it had pushed Einstein to explain the sense in which time does not exist in the theory of relativity—was cut short. Einstein doubtless better understood what Bergson had in mind when he read his book, *Duration and Simultaneity*. There he discovered that Bergson, starting from certain *a priori* statements, explicitly affirmed that the voyager imagined by Langevin would return to Earth having lived through exactly the same number of years as his companions remaining on Earth, and that it was therefore completely possible to continue to identify the duration experienced by each individual with a unique, universal time. Bergson's position amounted to claiming that the central element of Einstein's June 1905 article was false. Indeed, we have seen above that the change in the speed of the flow of time with the speed of an observer, and the twin paradox that it implies, constituted the central revolutionary conceptual contribution of Einstein's article.

Bergson, at any rate, never changed his mind. Indeed it seems that he thought until the end of his life that his book had "often been badly understood."[6] He regularly encountered Einstein in the '20s, in the setting of the committee on intellectual cooperation of the League of Nations. They were mutually appreciative of each other, but probably avoided speaking of their opposing conceptions of time. Einstein, in other company, had nevertheless commented on Bergson's philosophical concepts, frozen by his *a priori* assumptions, with a brief "may God pardon him." It is ironic that the first specific experimental tests of the twin paradox (high-precision measurements by Herbert E. Ives and G. R. Stilwell using atoms in motion, and less precise tests by Bruno Rossi and David B. Hall using cosmic rays) were published in

1941, the year of Bergson's death. However, well before this date, the great number of experimental facts either explained or successfully predicted by the theories of special and general relativity had already convinced the majority of physicists of the fundamental soundness of the Einsteinian conception of time.

The Princesse de Guermantes Listens to Einstein

Let us return once more to Einstein's visit to Paris in the spring of 1922, and let us no longer concern ourselves with Henri Bergson, but with his cousin, the writer Marcel Proust. Indeed, like his cousin Bergson, Proust had centered his entire work on the concept of time. However, in contrast to Bergson, his understanding of time, far from being opposed to that suggested by Einstein's theory, was remarkably similar. Some of Proust's readers, misled by the general title of his masterpiece, *À la recherche du temps perdu* (*In Search of Lost Time*), think that Proust's understanding is that of a time inexorably passing, which mankind can only nostalgically observe in its irreversible flight. In reality, however, this work is suffused with the idea that the passage of time is only an illusion, and moreover that, on rare occasions, a human being can have access to the "permanent and habitually hidden essence of things" and can feel that one's true self is "free from the order of time."

The entire development of *In Search of Lost Time* is directed toward its final volume, *Le Temps retrouvé* (*Time Regained*), where Proust unveils his philosophy of time, realized in an epiphany at a matinée party in the home of the Prince de Guermantes. There he describes the men, perched on the preceding years, as if they "were perched on living stilts, growing ceaselessly, sometimes higher than steeples." Otherwise stated, Proust has the vision of a reality where time is added to space, as a sort of vertical dimension, symbolized in the first part of the quote above by the image of *stilts*. In Proust's vision, the temporal flow is abolished, and the true self, "free from the order of time," is able, during certain privileged instants (realized for the narrator while looking at the steeples of Martinville, the trees of Balbec, etc.), to find joy in the unceasing adoration of reality. Proust's vision of an immobile time which is added to space as a new, vertical dimension is quite similar to the relativistic notion of space-time. Indeed, Proust was conscious of the relation between his ideas on time, and those issuing from the scientific work of Einstein.

In a letter from December 1921 to his physicist friend Armand de Guiche, Proust writes:[7]

> How I would love to speak to you about Einstein! Although it has indeed been written to me that I derive from him, or he from me, I do not understand a single word of his theories, not knowing algebra. And I doubt for my part that he has read my novels. It seems we have analogous ways of deforming Time. But I cannot figure it out for myself, because it is me, and we don't know each other, nor can I do so for him because he is a great mind in sciences that I am ignorant of, and from the very first line I am stopped by "signs" that I don't recognize.

In some of the preparatory manuscripts for *À l'ombre des jeunes filles en fleurs* (*In the Shadow of Young Girls in Flower*), he explicitly cites Einstein's name: "The faces of these young girls (very Einstein but do not say it, that would only confuse things) do not occupy in space a permanent size or form." Finally, in a 1922 letter to Benjamin Crémieux, speaking of an interval of time between the second trip to Balbec and the Guermantes matinée, whose length he is going to change, writes "Let's Einsteinize it," and he indicates that certain anachronisms appearing in the beginning of *In Search of Lost Time* took place "because of the flattened form that my beings take, from revolving in time."

With this context in mind, it is clear that Marcel Proust must have attentively followed the progress of Einstein's visit. He must have read the numerous articles which appeared in the Parisian press, either giving accounts of Einstein's lectures or attempting to explain his theories. Above all, however, I believe that Proust would have asked any friends present at the great public lecture of March 31st at the Collège de France to help him taste, as if he had been there, the unique atmosphere of that day. It is probable that his intimate friend, the physicist Armand de Guiche (to whom Proust had written concerning Einstein a few months before), attended Einstein's lecture. We do not have the complete list of the people present at this lecture, but it is striking to notice that among the small number of names explicitly cited can be found many of Proust's intimate friends. In particular, we find the names of Anna de Noailles, the Princess Edmond de Polignac, and, above all, of the Countess Henri Greffuhle.

This last friend, born Élisabeth de Caraman-Chimay, was connected to Proust, and to his work, in many ways: she was the mother-in-law of the Duke Armand de Guiche, she had been in frequent contact with Proust for many years, and, above all, she served as a model for one of the most important characters in Proust's work: the Princesse de Guermantes.[8] It is fascinating to imagine the possibility that Proust kept himself informed of Einstein's lecture on the notion of time through the intermediary of the Princesse de Guermantes!

The last year of Proust's life was 1922, and he dedicated his last bit of strength to finishing and perfecting *In Search of Lost Time*. According to Proust's devoted maid, Céleste Albaret,[9] it was in the beginning of the spring of 1922 that he again took up the formulation of the final phrase of *Time Regained*, concluding his description of the matinée of the Prince de Guermantes. Indeed, one afternoon, around 4 P.M., Proust called Céleste after waking up, to share some "important news": "Tonight, I wrote the word 'fin' ['the end']. [...] Now, I can die." Let us reread this final phrase, evocative of Einstein's space-time, which Proust had perhaps rewritten after having been informed of Einstein's notion of time through the intermediary of Armand de Guiche or of the Princesse de Guermantes:

> If at least there is enough time left to me to finish my work, I should not fail to mark it with the seal of this Time, of which the idea imposed itself on me with such strength today, and I will describe within it men as occupying a place, a very considerable place in contrast to the quite restrained one reserved for them in space, a place on the contrary prolonged without measure—since they simultaneously touch, like giants plunged into the years, such distant epochs, between which so many days have come into place—in Time.

A New "World": Space-Time

In the preceding chapter, we have seen that the essential element of Einstein's theory of relativity of June 1905 was the overthrow of the notion of time. Absolute, universal time, seeming to naturally coincide with the psychological duration experienced by everybody, was dethroned and replaced by a multiplicity of relative, individual times, disagreeing with each other, as shown by the twin paradox. The existence of this multiplicity of individual times, associated with particular physical processes, either of clocks or of biological organisms which measure or experience them, each irreconcilable with the other, had put into question the entire conceptual framework of Newtonian physics.

Significant progress in the physical understanding of this new concept of Einsteinian multiple times was accomplished by the mathematician Hermann Minkowski, who had actually been one of Einstein's professors at the Zurich

Polytechnic. In Cologne, on September 21, 1908, Minkowski gave a lecture before the twenty-fourth congress of German scientists and physicians entitled "Space and Time." This lecture marks the birth, from the point of view of *physics*, of a new "world," to use the word used by Minkowski to introduce the concept of space-time. His dramatic introduction has rightly remained famous:

> The views of space and time which I wish to lay before you have sprung from the soil of experimental physics, and therein lies their strength. These views are radically new. Henceforth Space by itself, and Time by itself, are doomed to fade away into shadows, and only a kind of union of the two will preserve an independent reality.

This union between space and time, which condenses the only meaningful reality of what had been described, before Einstein, by the separate concepts of space and time, was called by Minkowski "the world" or "the universe" (*die Welt*). Today it is called space-time. To understand deeply the essence of the conceptual revolution ushered in by the theory of relativity, one should familiarize oneself with the concept of space-time, and with its "chronogeometric" structure.

Recall that ordinary, Euclidean space, such as is taught in school, is a continuum with three dimensions (length, width, and height) whose structure is contained in the notion of *distance* between two points. Mathematically, the distance between two points is defined by a generalization of the Pythagorean theorem. More precisely, the square of the distance between two points is equal to the sum of the squares of the differences in length, width, and height between the two points considered.[10] The knowledge of the distance between any two points allows one to define all the other concepts of regular geometry. For example, one can define a straight line as being the shortest line joining two given points. One can also define the angle between two straight lines, crossing at a point *A*, as the length of the circular segment formed by the intersection of these two lines with a circle centered on *A* with a unit radius. A way of visualizing three-dimensional Euclidean space consists of representing, around each point in space, the locus of points which are separated from this point by a unit distance. In other words, one traces around each point a sphere of unit radius. The ensemble of all these spheres is a representation of the geometric structure of Euclidean space. See Figure 2. Having recalled the geometric structure of ordinary space, let's move on to the structure of space-time. First, what is a space-time point, as Minkowski would describe

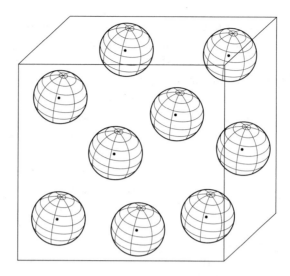

Figure 2. Geometric structure of Euclidean space.

it? It is an "event": something that happens at a precise point in space, at a determined instant. For example, this could be the collision between two particles, or, to take an example from everyday experience, a brief meeting between two people. To define an event one must, as for an appointment, indicate the point in space where it occurs and the instant when it occurs. One must therefore supply four numbers: three numbers (length, width, and height) serving to fix the event's spatial position, and the fourth (date) serving to fix its position in time. The necessity of giving four independent numbers to fix each space-time point indicates, in mathematical language, that space-time is a continuum of four dimensions. The four independent numbers which permit one to fix a point in the four-dimensional continuum are called, in mathematical jargon, the four coordinates of this point. One can thus consider the length, width, height, and date as the four coordinates of space-time.

Since it is difficult to picture such a four-dimensional continuum, let us consider the simpler case of a space-time with only three dimensions: two spatial dimensions and one time dimension. Such a three-dimensional space-time corresponds to the "world" of tiny insects living on a flat surface, for example, on the floor of a house. To define each space-time event for these insects, one must give three numbers, or three coordinates: the length and width fixing the spatial position of this event on the floor, and the date fixing its temporal position. One can then visualize this space-time by representing it in ordinary three-dimensional space: it suffices to identify the first two coordinates, length and width, of the insects' space-time with the length and width of ordinary three-dimensional space, and the third space-time coordinate, the date, with the height in ordinary space. We note in passing that we thus rediscover the image suggested by Proust in the final sentence of *Time Regained* quoted above, with time being conceived as a vertical dimension, symbolized by the image of stilts, added to the ordinary spatial dimensions.[11]

The notion of space-time helps to distill the essential physical novelty of the theory of relativity, so it is important to make a mental effort to get used to this concept, which represents a major change from the usual image that one has of reality. For example, one might train oneself by starting from the usual idea that the "world" of insects living on the floor is made of a succession of instants, each representing the state of the floor at each moment in time. Each instant describes the distribution on the floor, at the moment considered, of all the insects living there. This spatial distribution, at a definite moment, can be completely described by a photograph, an instantaneous snapshot of the floor's surface. In this way, the three-dimensional space-time of the insects living on the floor is obtained by *piling up*, vertically, the continuing series of these snapshots, with each one representing the spatial state at a moment in time—like a deck of cards, made here from stacking cards representing individual moments. The height of each snapshot in this pile is proportional to the date corresponding to this snapshot.

For each insect there is a spot on each photograph in the pile, and at each instant of time there is a spot for each insect. The experience of each insect thus defines a continuous succession of spots, which traces a tube (or rather a thick line) in space-time. These are the *stilts* in Proust's imagery. If the insect remains at rest on the floor, its space-time tube (or world-tube for Minkowski) rises vertically, that is, orthogonally to the horizontal directions representing the space in which the insects live. On the other hand, if the insect is moving, its space-time tube will be inclined with respect to the vertical. The faster it moves, the more the tube inclines. If we consider an insect who, like a race-car driver, runs quickly in a circle, its space-time tube becomes a helix with a vertical axis. I shall leave it to the reader to train his or her imagination by considering the space-time figures formed by more complicated insect configurations, then the figure formed by a collision between two insects, up to the figure formed by a pitched battle between two tribes of insects, passing through a great number of entomological choreographies.

The reader will have perhaps recognized that these three-dimensional space-time representations are a generalization of the train diagrams used once upon a time (at least in Europe) to resolve problems in train crossings. For example, consider two trains moving in opposite directions along the same route, leaving from two stations at different instants and at different speeds. We want to determine where and when the two trains are going to cross. A simple way to solve this problem is to represent the history of displacement of the two trains in a two-dimensional diagram, where the horizontal direction represents the distance along the railway, and the vertical direction represents time. This diagram is an example of a two-dimensional space-time, with one spatial dimension (length) and one time dimension. In this space-time, each train traces a continuous line, made from segments of straight lines whose inclination with respect to the vertical depends on the speed of the train (with a train at rest tracing a vertical line). The event corresponding to the crossing of the two trains is represented by the space-time point which is the intersection of the lines traced by the two trains: the horizontal projection of this point, in other words, its first coordinate, measures the length along the track of the location where the trains cross in space, while its vertical projection, the second coordinate, gives the time at which the crossing takes place.

Methodological Interlude on the Notion of "Real"

The skeptical reader may at this point be thinking that it is very well to use mathematical jargon to describe space and time together in a four-dimensional continuum, but that this is nothing but an artificial construction, like the union of a fish and a bicycle. This same skeptical reader might think that this construction does not take into account the essential difference between time and space: the fact that, at every instant, only the "now"—the present instant—is "real." The past no longer exists, and the future does not yet exist, and thus neither has any immediate reality. He would thus say to himself that if one represented reality in space-time, one must add the information that only one horizontal slice of this space-time is "real" at each instant. For example, one might imagine that among the pile of snapshots representing the world of insects living on the floor, only one snapshot is illuminated, that corresponding to the present instant, and that this state of illumination propagates continually from bottom to top through the pile of snapshots. Such an image would have pleased Bergson, for it permits one to keep the common notion of duration, experienced in its eternal motion. But, in fact, the point of view of this book is that the theory of relativity, and notably the numerous experimental tests of the twin paradox, obliges us to reevaluate the ordinary notion of passing time.[12]

Before going any further, we shall explain the author's point of view concerning the link between the scientific representation of "things" and the notion of "reality." This is, of course, a centuries-old philosophical question, which must be addressed subtly. The point of view adopted in this book will essentially be that introduced by the philosopher Immanuel Kant.[13] It consists essentially in saying that the physical universe only acquires an objective existence, as a suitably scientific object, within a logical and mathematical framework which must be imposed *a priori* by the human mind. Kant's thought has sometimes been caricatured by saying that he had affirmed, among other things, that the development of science necessitated the *a priori* postulation of the existence of ordinary Euclidean space. This would be in contradiction to Einstein's theory of general relativity, which demonstrated that space was not Euclidean. In fact, Kant's thinking is much more subtle than that. Moreover, the study of Kant which Einstein made in his youth probably helped him open his mind and develop his own scientific philosophy, whose essential feature was the liberty of a theorist to invent a new scientific framework in order to explain the facts.

In sum, the attitude taken here will consist in taking seriously what science tells us. In other words, we take the position that it is science which defines what is real. As Kant said:

> Hitherto it has been assumed that all our knowledge must conform to objects.[...] Let us try to see whether we may not have more success in the tasks of metaphysics, by assuming instead that objects must conform to our knowledge.

Of course, we shall only adopt this attitude with regard to those parts of our scientific knowledge which have been confirmed by a very large number of experimental tests. This is certainly the case for the special theory of relativity, the general theory of relativity, and the quantum theory (both relativistic and nonrelativistic).

The Space-Time Block

In accord with the methodology we have just described, we now describe the new notion of reality defined by the theory of relativity. As Minkowski was the first to explain, relativistic reality must be thought of in space-time, in a four-dimensional block where all temporal flow is banned. An example of such a space-time block is, as we discussed above, the pile of snapshots representing the history of insect configurations living on the floor. Indeed, there is nothing in the formalism of special relativity corresponding to the idea of a "now," of a privileged instant describing the present. Moreover, there is no way to define a succession of privileged "horizontal slices" which could correspond to the common idea of universal time. This is the foundation of what fascinated the journalists when, in the '20s, they began to take interest in Einstein and his theories. It is also what bothered Bergson, who cornered himself into imagining that twins in relative motion both age at the same rate, in order to continue believing in his concept of universal duration.

Today, after many experimental confirmations of the reality of the twin paradox (see Chapter 1), we no longer have this choice: the multiplicity of equally valid temporal threadings of space-time is a reality. We are cornered by experimental facts into accepting the necessity of thinking in terms of a space-time block, without any sort of fixed temporal flow.

A musical analogy may be useful. Space-time describes the collective history of reality *sub specie aeternitatis*, just as a score describes the whole of a musical work. The score exists in a "motionless" way, even though it describes something which is generally understood by the human mind in the form of a temporal flow. The reader will perhaps think that this comparison rather suggests that a "motionless" space-time is no longer capable of accounting for real motion, in the same way that the picture of a musical work as a block would not be able to correctly convey what music is.

Before jumping to such a conclusion, let us ask one of the greatest musicians of all time what he has to say on the question. Indeed, Mozart gave an interesting description of the way in which a work would form, nearly spontaneously, in his mind. First, he described how the musical ideas came to him in great number, then, after a process of selection, they hooked together between themselves to form a coherent work. He continued by saying:

> The work is thus completed within my skull, or really just as well, even if it is a long piece, and I can embrace the whole all at once like a painting or a statue. In my imagination, I do not hear the work as it unfolds, as it must play out, but I hold it all as one block, so to speak. What a delight! Invention, elaboration, all of that only happens within me as a magnificent and grandiose dream, and when I succeed in thus "super-hearing" the entire assembly, it is the best moment.

As this quote shows, a great musician can transcend the usual fashion in which mere mortals hear music with a "super-hearing" of a "block," outside of any temporal unfolding. The structure of the theory of relativity suggests that if one could break through the thermodynamic and biological constraints which condition us, in everyday life, to experience reality in the form of a temporal flow, one could, by analogy, "super-live" our life as a block, a block making up a small part of Minkowski's four-dimensional space-time.

The World's Checkerboard

We have recalled above that ordinary Euclidean space is defined by giving *two structures*: (i) it is a three-dimensional continuum (with points that are fixed by giving three continuous coordinates), and (ii) the distance between two points is defined by a simple formula, derived from the Pythagorean theorem. All other structures of Euclidean space can be deduced from these two fundamental properties.[14] We have already indicated what the analog of the first structure is for the space-time of special relativity: it is a four-dimensional continuum, whose points (called *events*) are fixed by four continuous coordinates: length, width, height, and date.

It remains for us to describe the analog of the notion of distance between two events. This analog was introduced by Henri Poincaré, and is called the *space-time interval* between two events. It is defined by a mathematical formula nearly as simple as that giving the distance between two points in ordinary space. This formula rests on a generalization of the Pythagorean theorem which seems innocent enough, but which is rich in remarkable physical consequences: the squared interval between two events *A* and *B* is equal to the *sum* of the squares of the differences in length, width, and height of *A* and *B*, *minus* the square of the difference in date between *A* and *B* (multiplied by the speed of light).[15] In contrast to the usual case of Euclidean space where the distance squared between two points is always given by a *sum* of squares, all taken with the same plus sign, one sees here that the squared interval contains four terms: three squares preceded by a plus sign, and a fourth square preceded by a minus sign. This final minus sign is essential, and has a number of physical consequences.

Note first that, because of the presence of this minus sign, the squared interval between two events is not always positive (despite the fact that it is designated as a square). It can be positive, negative, or zero. When the squared interval between two events is zero, it implies that these two events can be connected by a ray of light (see the notes to Chapter 1 at the end of this book). When it is negative, it means that the two events can be connected by the trajectory of a massive object (an atom, or an observer) moving at a speed smaller than that of light. In this case, the squared interval between the two events is equal, after changing the sign and dividing by the square of the speed of light, to the square of the time lapse experienced by this atom or this observer while passing, at constant speed, from one event to the other. Finally, in the case where the squared interval between two events is positive, it means that there exists an observer for which these two events are simultaneous and separated by a spatial distance whose square is equal to this squared interval.

In sum, we see that the squared interval between two events essentially measures, according to its sign, a distance squared, or oppositely a duration squared (multiplied by the speed of light squared). We see also that the speed of light plays the role of conversion factor between duration and distance. To simplify, it is convenient to use units where there is no need to introduce this factor explicitly. To do this, it suffices to measure, for example, duration in seconds, and distances in *light-seconds*. We recall that a light-second is the name given to the distance traveled by light in one second (in the same way that a light-year is the distance traveled by light in one year). A light-second is thus equal to 300,000 kilometers. In such units, the speed of light is 1 (that is, one light-second per second). In the following, we will generally suppose that we are using such units.

The notion of the squared interval between two events defines what one may call the *chronogeometry* (or, if one prefers, the *geochronometry*) of space-time, that is to say the generalization of the geometry of ordinary space, such as is defined by the notion of distance between two points. The geometry of space could be visualized by representing, around each point P, the ensemble of points at unit distance from P, which is a sphere. In the same way, one may visualize the chronogeometry of space-time by representing around each event P the ensemble of events separated from P by a unit squared interval. As the squared interval between two events can be positive, negative, or zero, we see that a complete representation of the chronogeometry of space-time will consist of tracing, around each event P, three ensembles of events: (i) the ensemble of events separated from P by a squared interval equal to *plus* one, (ii) the ensemble of events separated from P by a squared interval equal to *minus* one, and (iii) the ensemble of events separated from P by a squared interval *equal* to zero.

Figure 3. Chronogeometric structure of the Poincaré-Minkowski space-time.

These ensembles of events are no longer spheres, as was the case for Euclidean geometry. The reader will find in Figure 3 the representations of the different ensembles (i), (ii), and (iii). Note that ensemble (iii) is a double cone, made up two cones connected at their points (with one cone opening toward the "top" of space-time, which is conventionally called the future, while the other cone opens toward the "bottom" of space-time, called the past). As this cone represents the events which are connected to the event P by a ray of light, it is called the *light-cone*. The ensemble (i) has the form of an hourglass (in other words, it resembles two cones joined at their apexes, and then carved so that there is an open throat through which sand can flow). The ensemble (ii) is made of two separate sections: one is contained above the top part (toward the future) of the light-cone, and the other is contained below the lower part (toward the past).

The figure representing the chronogeometry of space-time (Figure 3) forms what might be called the world's checkerboard. The world, in the sense of Minkowski, denotes space-time, and the checkerboard visualizes the permitted ways of connecting the squares of the checkerboard, in other words, the possible connections between the various events in space-time. For example, the light-cone indicates a connection between two events made by the exchange of a ray of light. It is amusing to note that this new "checkerboard" is formed from figures which resemble hourglasses. The flow of time

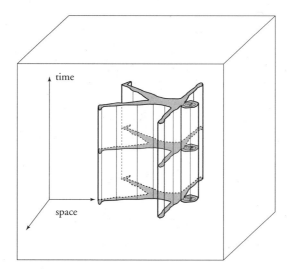

Figure 4. The space-time tube traced by a man.

is abolished in space-time, but each hourglass reminds us that the structure embedded in this atemporal world still produces the illusion of irreversible flow. Heraclitus might have appreciated this picture of the world's checkerboard, he who imagined time as a child who plays at checkers, as shown in the quote which opens this chapter.

The checkerboard of Minkowski's world is empty. It represents the spatio-temporal canvas which serves as a frame of existence for matter and its interactions. To give observational meaning to the chronogeometry (or space-time structure) of this world, one must populate it with objects capable of probing this structure. Let us recall that, as in the example given above of the world of insects living on the floor, an object having a significant lifetime, like an insect, traces a tube which extends from the bottom to the top of space-time. A person's life may also be visualized by such a space-time tube (see Figure 4). This tube corresponds to the stilts mentioned previously by Proust. We note as well that Proust's intuition was correct: this tube occupies a place much more considerable in time than in space. Indeed, by measuring, as we have discussed, durations in seconds and distances in light-seconds, this tube has a (temporal) height of some *billions* of seconds, while its (spatial) width is on the order of *billionths* of a light-second. In other words, the ratio between the height and the width is on the order of a billion billions. In the limit where one considers an object of very small spatial extent, let's say an atom

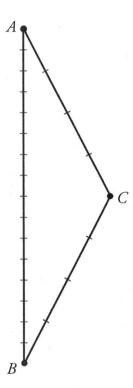

Figure 5. Space-time triangle and the twin paradox.

or an elementary particle, its space-time tube is reduced to a simple line of negligible width. This space-time line crosses space-time from the bottom to the top and only terminates when the particle appears or disappears (see an example in Figure 3).

Finally, it is important to see what becomes of the twin paradox in space-time. The essence of this phenomenon can be illustrated by considering the chronogeometry of a *space-time triangle*. Let it be the space-time triangle *ABC* (see Figure 5). The sides of this triangle represent the space-time lines traced by two twin clocks (of identical construction). The first clock ticks along side *AC* (which lies in the "direction" of time, corresponding to a negative squared interval), while the second clock ticks first along side *AB*, then along side *BC* (these two sides also lying in the direction of time). The fact that sides *AB* and *BC* are inclined with respect to side *AC* corresponds, in the usual decomposition of space-time into components of space and time, to saying that the second clock, which was originally coincident with the first (at event *A*), first moves away at constant speed, and then returns at a constant speed and meets up with the first clock at space-time point *C*.

The theory of relativity tells us that the duration experienced by one clock is given[16] simply by the total interval along the space-time line traced by the clock. Therefore the first clock will tick out a duration equal to the length (in Minkowski's meaning) of side *AC*, while the second clock will tick out the sum of the lengths of the other two sides of the triangle: *AB* + *BC*. The chronogeometry of space-time, in other words, the specific form of the hourglasses defining the world's checkerboard, tells us that the sum of the sides of the triangle, *AB* + *BC*, is *shorter* (in interval) than the third side *AC*. Therefore the number of ticks of the second clock, which went from *A* to *B* and then returned from *B* to *C*, is smaller than the number of ticks of the first clock which took the direct space-time route from *A* to *C*. This space-time triangle inequality[17] has behavior contrary to that of the triangle inequality in ordinary Euclidean space, for which the sum of two sides is always longer than the third side. This difference is due to the particular form of the chronogeometry of space-time, where the Pythagorean theorem contains a minus sign for the squares of the sides of a triangle which lie in the direction of time.

Independently of this question of whether a direct path is longer or shorter than an indirect path, the essential point remains that the very existence of a geometric interpretation of the duration measured by clocks shows the *nonexistence* of an absolute, universal duration. For Newton, absolute time ticked along in a universal and regular fashion in all of space, for all clocks and for all organisms sensitive to the passage of time. For Einstein, there is no longer a temporal flow, and each clock sees its own time, which simply measures the "length" of the line it traces in space-time. Thus there are as many times as there are clocks.

A Tenacious Illusion

In this chapter, as well as in the preceding one, we have continually asserted that Einstein's theory of relativity overturned the common notion of time and, in particular, showed the unreal (in the Kantian sense) character of temporal flow.[18] This unreality (in the sense of the objectively real) intrigued Einstein for a long time. In spite of the numerous articles written by journalists who mention the problem associated to the reality of temporal flow, there are, in fact, few documents where Einstein expresses himself directly concerning this problem. He did state his views in conversation with the philosopher Rudolph Carnap, in writing within the work *Relativity and the Problem of Space*,[19] and above all in the many letters that he wrote to his most intimate friend, Michele Besso, the very same friend who had been crucial in helping him to understand that the notion of time must be modified to satisfy the principle of relativity. Let us quote some of the comments written by Einstein to his friend in order to correct what Besso had said (since Besso had not, apparently, understood what relativity implied for time), or to clarify his position:

> The "now" [...] is eliminated from the conceptual construction of the objective world. (This is what was bothering Bergson.)

> ...you do not take seriously the four-dimensionality of relativity, but you consider the present as the only reality. That which you call the "universes" are in physical language "spatial slices," to which the theory of relativity (even in its restricted form) denies any objective reality.

> You cannot get used to the idea that subjective time with its "now" should not have any objective significance. See Bergson!

But the most touching comment written by Einstein on this problem is contained in the letter of condolence which Einstein wrote to Besso's son and sister after the sudden death of his friend (on March 15, 1955). We note that this beautiful letter was written on March 21, 1955, less than one month before Einstein's own death:

> Now he has departed from this strange world a little ahead of me. That signifies nothing. For us, physicists in the soul,[20] the distinction between past, present, and future is only a stubbornly persistent illusion.

Lorentz, Poincaré, Einstein, Minkowski, and Special Relativity

Recent years have seen an entire literature[21] flourish seeking to "rehabilitate" the contributions of the French mathematician Henri Poincaré to the (special) theory of relativity. The author of this book is an unconditional admirer of the mathematical genius of Poincaré, and also an admirer of his popularization books (notably *The Value of Science*) which he warmly recommends to the reader since, one century after their publication, they are of a conceptual richness rarely equaled. Nevertheless, after having read the pro-Poincaré (and often absurdly anti-Einstein) literature, and having reread in detail Poincaré's own writings, the author has come to the conclusion that the standard "cartoon" vision of the history of relativity is ultimately correct: Poincaré went significantly farther than Lorentz, and he discovered (in 1905) an important part of the mathematical structure supporting the special theory of relativity. He did not, however, succeed in making the major conceptual "step," concerning the nature of time, with which Einstein founded his theory in June 1905. The second major conceptual step, that of proposing the replacement of the separate categories of space and time with the new physical category of space-time, is likewise more properly attributed to Hermann Minkowski, and not to Poincaré.

To discuss all of these questions in detail would take up a large amount of space, and we shall content ourselves here with some brief remarks highlighting the essential points.[22] Before entering into details, let us emphasize that the critiques formulated below do not diminish Poincaré's contributions in any way. They simply aim to characterize the major differences between the ideas of Poincaré and those of Einstein. If it had been a question of giving, before Poincaré's premature death in 1912, a Nobel prize in physics for the discovery of the special theory of relativity, it would not seem out of place to imagine a shared prize between Lorentz, Poincaré, and Einstein. Each of them had indeed brought decisive contributions to the final formalism of this theory.

Three of the crucial points discussed by Poincaré's flatterers are: (i) the procedure to synchronize clocks in motion by the exchange of electromagnetic signals (discussed by Poincaré in 1900 and 1904), (ii) the fact that Poincaré had spoken (in September 1904) of a "principle of relativity," and had placed it at the same rank as other important physical principles, and (iii) the fact that he had introduced the mathematical structure of space-time (in July 1905). We have already discussed the first point in the preceding chapter, and showed that a close reading of Poincaré's writings show that he had never thought of, nor even technically derived, the effect of time dilation, which is the essential conceptual novelty of Einstein's revolution. Concerning point (ii), two other facts indicate that Poincaré had a totally different approach from that of Einstein with regards to the statement of an eventual "principle of relativity." First, in an article published in 1908, Poincaré mentions Kaufmann's latest experiments on the dynamics of electrons at high speed, which contradicted the predictions for relativistic dynamics of Lorentz (and Einstein), and he immediately abandons his confidence in the validity of the principle of relativity:

> [These experiments] *have shown Abraham's theory to be correct.* The Principle of Relativity thus must not have the rigorous value that one was tempted to attribute to it; one should no longer have any reason to believe that positive electrons are deprived of real mass like negative electrons.

Note that the emphasis is Poincaré's, and that Abraham's theory was contemporary with that of Lorentz, and did not satisfy the principle of relativity. I have also left the last part of Poincaré's statement, since even though it will seem obscure to the modern reader, it shows that Poincaré was thinking within a particular framework of thought, which was completely different from that of Einstein. Roughly stated, Poincaré's framework was shared by the other contemporaries of Einstein, such as Hendrik Lorentz, Max Abraham, Emil Cohn, or Paul Langevin. It consisted in founding the electrodynamics of moving bodies on a choice of particular hypotheses concerning the microscopic constitution of matter (and ultimately of the ether). From this point of view, any principle of relativity appeared not so much as a starting postulate, but rather as a result to be demonstrated, starting from detailed hypotheses concerning matter and the forces acting on it.

The framework of Einstein's thinking was completely different. One could even say that it was the *opposite* of that of his contemporaries. In fact, Einstein starts from the result (the principle of relativity) that the others tried to deduce from hypotheses about matter, and he places it as a postulate, as a starting point and a tool to then derive some general results concerning the structure of matter. In sum, as remarked by Einstein himself, his contribution was to reverse the problem. This approach, consisting of assuming a symmetry principle as a basic postulate and then deriving general results on the structure of matter and its interactions, is very modern, and has proven to be quite fruitful in many different domains of twentieth-century physics. When Einstein introduced it, however, it was totally new, and it shocked some physicists (like Lorentz) who thought that Einstein had cheated by assuming as a principle what he needed to prove.[23]

Let us show with two precise examples the difference in approach between Einstein and Poincaré. In 1907, speaking of the same experiments performed by Kaufmann, which Poincaré had interpreted as falsifying the principle of relativity, Einstein proposed a totally different interpretation:

Taking account of the difficulties of experiment, one should be inclined to consider the agreement [between Kaufmann's results and the predictions of the theory of relativity] as sufficient. The discrepancies observed are nevertheless systematic, and exceed in a significant fashion the error bar of Kaufmann's experiment. [...] Do these systematic discrepancies have as their origin a source of error as yet unknown, or the fact that the foundations of the theory of relativity do not correspond to reality? This can only be settled with certainty when one produces more diversified observational results.

It must still be mentioned that the theories of electron motion proposed by Abraham and by Bucherer give curves that are clearly closer to the observed curve [of Kaufmann's experiment] than the curve obtained by the theory of relativity. But the verisimilitude that one may attribute to these theories is in my opinion quite meager, because their fundamental hypotheses concerning the mass of an electron in motion are not suggested by theoretical systems which encompass larger ensembles of independent phenomena.

A second example where one sees the difference of perspective between Einstein and Poincaré in action is given by the only interaction they ever had concerning the theory of relativity. Einstein and Poincaré only met once, in 1911, at the first international Solvay council in Brussels. This conference was dedicated not to relativity, but to the very newly developed quantum theory. Nevertheless, Maurice de Broglie reports that one day:

As Einstein was explaining his ideas [on relativity], Poincaré asked him: "Which mechanics do you adopt in your reasoning?" Einstein responded: "No mechanics," which seemed to surprise his interlocutor.

To conclude, let us state that although the first discovery of the mathematical structure of the space-time of special relativity is due to Poincaré's great article of July 1905, Poincaré (in contrast to Minkowski) had never believed that this structure could really be important for physics. This appears clearly in the final passage that Poincaré wrote on the question, some months before his death. This text[24] contains some passages which, when quoted out of context, seem to show Poincaré passionately arguing for the physical interest of four-dimensional space-time:

Everything occurs as if time was a fourth dimension of space; [...] it is essential to note that in the new conception, space and time are no longer two entirely distinct entities, which one can envisage separately, but two parts which are as if tightly laced together, in such a way that one can no longer separate them.

However, in reality this passage of Poincaré's only presents the new conception, or as Poincaré would prefer to say, new convention of "certain physicists" (without ever citing either Einstein[25] or Minkowski) in order to better distance himself from it. Indeed, the final paragraph of this passage is:

> What will our position be, confronted with these new ideas? Will we be forced to modify our conclusions [concerning the liberty that one has to adopt the conventions that one finds useful]? Certainly not: we have adopted a convention because it seemed useful to us, and we have said that nothing could force us to abandon it. Today, some physicists want to adopt a new convention. It is not that they are forced to; they judge this new convention to be more useful, that is all; and those who are not of the same opinion may legitimately keep the former convention in order to not trouble their old habits. I think, between us, that this is what they shall do for a long time.

One here sees (in contrast with the eminently fruitful character of Einstein's approach) the sterility of Poincaré's scientific philosophy: complete and utter "conventionality." It is quite probably Poincaré's epistemological attitude, combined with his skeptical idealism, his mathematician's vision of physical reality, and his conservatism, which stopped him from taking seriously, and developing as a physicist, the space-time structure which he was the first to discover.

Ephemeral Matter

The fruitful nature of Einstein's "new conception," consisting of the postulation of the principle of relativity as a *constituent symmetry of reality*, and the subsequent deduction of general properties of matter and of its interactions, became apparent quite rapidly. A few months after his article of June 1905, Einstein understood that the new *relativistic symmetry* implied the following remarkable conclusion: mass is a measure of the energy contained in a body. This is reflected in the most famous equation in twentieth-century physics: $E = mc^2$. Here, m is the mass of a body, and this equation associates to this mass an energy E, which represents the energy contained within the body.[26] This equation is fascinating for its simplicity and its depth. It has definitively changed our ideas about matter.

For Newton, the mass of a body was thought of as its "quantity of matter." On the other hand, for centuries matter had been conceived as an eternal substance, which was conserved in all of its transformations, even if its exterior appearance changed or it recombined into a new form. This is reflected in the famous dictum of Antoine Lavoisier: "nothing is destroyed, nothing is created, all is transformed." He asserted that the mass remained invariant under all the transformations of matter. Lavoisier had experimentally confirmed this conservation of mass through various chemical reactions, recombining matter into new forms.

Ever since Einstein and the equation $E = mc^2$—or rather the equation in its reciprocal form $m = E/c^2$—mass, and thus matter, has lost its permanence. If a body gains energy, for example, by heating, its mass grows. On the other hand, if a body loses energy, for example, in the form of electromagnetic or other radiation, its mass diminishes. In his article of September 1905, where he first introduces the relation $E = mc^2$, Einstein mentions the eventual verification of this *equivalence between mass and energy* in the case of radium, which had been discovered seven years earlier by Pierre and Marie Curie. Radium posed a problem for classical physics since it seemed to violate the principle of conservation of energy. It continually emitted energetic rays. The Curies showed, by putting radium in a calorimeter, that the quantity of heat (and thus of energy) incessantly created by the radium was quite considerable. Nevertheless, this ceaseless energetic activity did not seem to affect the radium at all, which seemed to subsist without alteration for years, and thus seemed to be an inexhaustible source of energy. Einstein predicted that on the contrary this emission of energy must necessarily be made at the expense of the mass of the emitting body. Therefore it must affect the radium in a measurable way. Nevertheless, in the usual units, the speed of light is expressed by a very large number. Its square is thus a gigantic number, and Einstein's relation, $m = E/c^2$, predicts that the loss of mass corresponding to the energy lost by radium is absolutely minuscule.

Some years later, Paul Langevin (who, independently of Einstein, had foreseen the existence of the relation $E = mc^2$) suggested that it would be easier to experimentally verify the mass-energy equivalence in the case of nuclear reactions, where certain atomic nuclei are recombined or disintegrate into other nuclei. The first experimental verification of the relation $E = mc^2$ was obtained by Cockcroft and Walton in 1932, who bombarded lithium 7 nuclei with protons, and considered the events where this collision caused the fission of the bombarded nuclei into two nuclei of helium 4. Since we happen to be speaking here of nuclear reactions, let us mention that, contrary to legend, the relation $E = mc^2$ in no way aided the discovery of the possibility of an atomic (or rather, nuclear) bomb, nor did it aid in its conception or realization. In fact, the most direct way to explain the origin of the enormous energy released by an atomic bomb founded on fission, or by a nuclear fission reactor, consists in saying that it is essentially an electrical energy of repulsion between the positively charged protons of the fissile nuclei considered. A simple calculation, based on Coulomb's law for electric force (known since the 1790s), and on the radial size of the atomic nucleus, gives a good estimate of the energy released by the fission of a nucleus, without it ever being necessary to appeal to the equivalence between mass and energy.

Outside of the numerous scientific consequences of the relation $E = mc^2$, and its implied mass-energy equivalence, the most fascinating consequence of this relation is the major conceptual slide it makes from permanent matter to *ephemeral matter*. Ever since the Greeks, and the concept of indivisible atoms which they introduced, matter was thought of within the paradigm of a permanent substance underlying reality. Since Einstein, matter has lost its substantial permanence. At the very least, it can no longer be connected to a particular, substantial form. It can transform from one substance to another, or even be totally disintegrated into luminous energy or any other form of radiation previously considered as being immaterial.[27] A particularly striking example of the ephemeral character of matter is the possibility, opened up by mass-energy equivalence, of the total disintegration of an atom of positronium, made up of an electron and a positron, into electromagnetic radiation. The initial state is a system made of two material particles which are separately perfectly stable: an ordinary, negatively charged electron (also called a negatron), and an electron of positive charge (or positron). This initial "material" state spontaneously disintegrates when the negatron gets too close (in classical terms) to the positron, and is transformed into a final "immaterial" state, made up entirely of electromagnetic radiation (more precisely of two quanta of light; see Chapter 5). This process was observed and studied in detail during the late 1940s, and one was able to then verify with precision the validity of the relation $E = mc^2$. Even more striking is the inverse process, which has also been observed many times: one may start with an initial "immaterial" state, consisting uniquely of electromagnetic radiation (for example, two quanta of light with sufficiently high energy and headed for a collision), and this initial state leads to the creation of two (or more) "material" particles. The reader will perhaps think that these are exceptional cases, having no concrete, practical implications for the ordinary matter which surrounds us. Far from it! Modern cosmology suggests, on the contrary, that all of the matter which surrounds us, and of which we are made, did not exist during the first stages of expansion of the universe. Instead, it was created from the energy stored in a continuous field (similar to the electromagnetic field) initially present in all of space.

A Profound Simplification of the Basic Categories of Reality

To summarize the conceptual revolutions ushered in by the two works we have discussed, completed by Einstein in 1905, one could say that before 1905 physical reality was described by way of four quite distinct categories: *space, time, force,* and *matter.* The paper from June 1905 led to the disappearance of the two categories of space and time, to be replaced by a new fundamental category of reality: space-time. As for the work of September 1905, it began to erase the traditional distinction between force and matter. Indeed, on one hand, matter was traditionally associated with mass, since the latter was thought of as the "quantity of matter." On the other hand, to each force there necessarily corresponds an *energy of interaction* between material objects. For example, the electric force is connected to the electric energy of attraction or repulsion between electric charges. However, the relation $E = mc^2$ associates to each energy (and in particular to any interaction energy) a certain contribution to the mass. Thus, translated into the terms of the former categories, this relation associates to each force (of interaction) a certain (quantity of) matter. In modern terms, it is preferable to forget the old separate categories of force and matter, and replace them with a new fundamental category: *mass-energy.*

To conclude, we note that the new category of mass-energy satisfies the fundamental requirement of permanence that Lavoisier had previously imagined to hold for mass alone. In a relativistic system, the sum of the masses of material particles is not separately conserved, nor is the interaction energy, or more generally, the energy stored in various fields (such as the electromagnetic field). However, the total mass-energy (which represents the sum of the energy stored in the form of material masses, the kinetic energy of matter, and the energy stored by continuous fields) is conserved through all possible transformations. This includes those transformations where, for example, all matter disappears and is transformed into the radiation energy of various fields.

3

Elastic Space-Time

Probably the greatest scientific discovery that was ever made.
—P. A. M. Dirac, speaking of the general theory of relativity

Newton Unthroned

London, England, Thursday, November 6, 1919

The atmosphere was quite tense on that Thursday, the 6th of November, 1919, at the seat of that most venerable British scientific institution, the Royal Society. All of the leading British scientists had come that day to Burlington House, where a joint session of the Royal Society and the Royal Astronomical Society was to be held. The president of the Royal Society, Sir Joseph John Thomson (famous for his discovery of the first elementary particle, the electron), opened the session. He recalled that the aim of this joint meeting was to present the results obtained by two British astronomical expeditions which had gone to observe the total eclipse of the Sun on May 29, 1919. The observations made by these expeditions consisted of photographs (taken by day) of the field of stars around the Sun at the moment when the eclipse was total and allowed the stars to be seen. These photographic plates were then compared to photographs (taken at night) of the same field of stars, at a moment when the Sun was far from the field. The aim of these observations was to confirm or invalidate a prediction made by Einstein in November 1915, which was a consequence of his new conception of gravity based on a generalization of the theory of relativity of 1905.

Everyone in the room knew that the stakes were immense for these observational results. It was a battle on the summit between Newton and Einstein. In 1686, Newton had founded a mathematically precise theory of gravity, which was based on his concepts of absolute space, absolute time, and a gravitational force which grew weaker with the square of the distance between

57

two objects. This theory did not say anything precise *a priori* on the fashion in which gravitation might affect light. According to the most common beliefs of the early twentieth century, where light was described as an electromagnetic wave, Newton's theoretical framework seemed to predict that gravitation should not have any influence on light. In particular, rays of light passing close to the edge of the Sun (as is the case during an eclipse for the light from the stars which appear around the Sun) should not be affected by its gravitational field. If, on the other hand, one returned to the conception held by Newton himself, according to which light was made of tiny material projectiles, one could predict that the Sun should deflect the rays of light passing close to its edge inwards by an angle equal to 0.875 arc-seconds. As for Einstein's new theory, it made a precise prediction:[1] the Sun should deflect light rays passing close to its edge inwards by an angle equal to 1.75 arc-seconds, exactly *double* the Newtonian value (in the second interpretation given above).

Newton's portrait, overlooking the venerable meeting hall of the Royal Society, gave a particular solemnity to this session. Newton was not only the greatest British scientist ever to have lived, but he had also been president of the Royal Society for many years. Depending on the nature of the results which were soon to be announced, either the centuries-old Newtonian conception of absolute space and absolute time would be confirmed, or Newton would be unthroned in favor of Einstein and his new conception of a curved space-time; a conception which seemed rather obscure to most of the scientists assembled that day at Burlington House. We note as well that Einstein was a German scientist,[2] and that he had developed his theory in Berlin, during the first World War. The sense of fair-play shown by the British scientific community (which rejected the most famous British scientific theory in favor of a theory developed in an enemy country) is remarkable.

To help us imagine the exceptional ambiance of this session of November 6, 1919, let's listen to the testimony of an eye-witness, the mathematician and logician Alfred North Whitehead:

> The whole atmosphere of tense interest was exactly that of the Greek drama. We were the chorus commenting on the decree of destiny as disclosed in the development of a supreme incident. There was dramatic quality in the very staging: the traditional ceremonial, and in the background the picture of Newton to remind us that the greatest of scientific generalizations was now, after more than two centuries, to receive its first modification. Nor was the personal interest wanting: a great adventure in thought had at length come

safe to shore.

Sir J. J. Thomson gave the floor to the first speaker: the royal astronomer, Sir Frank Dyson, who had had the necessary determination to approve and organize (after a suggestion by the physicist Arthur Eddington) two delicate scientific expeditions during a particularly murderous war. Dyson described the aim of the expeditions, their equipment, and the intricate procedures of analysis applied to the observational data. He concluded by declaring:

> After a careful study of the plates, I am prepared to say that there can be no doubt that they confirm Einstein's prediction. A very definite result has been obtained that light is deflected in accordance with Einstein's law of gravitation.

Following this, he left the floor to the scientists who were responsible for each of the expeditions: first, A. C. D. Crommelin for the expedition which had gone to observe the eclipse at Sobral, Brazil, then Arthur Eddington for the expedition which had gone to the isle of Principe, along the African coast. The two results for the deflection of light passing along the edge of the Sun (about 2 arc-seconds for Sobral, and about 1.6 arc-seconds for Principe) confirmed Einstein's prediction, and ruled out the two conceivable predictions of the Newtonian framework of physics. The error bars for these results were rather large (6% for Sobral and nearly 20% for Principe), and probably optimistic, but the coincidence with Einstein's prediction and the disagreement with the Newtonian predictions succeeded in convincing (nearly) everyone during the discussion following the presentation of the observational results. The only discordant voice which was raised was that of the physicist Ludwig Silberstein who pleaded for caution. In a dramatic gesture, he took as a witness Newton's portrait, while exclaiming: "We owe it to that great man to proceed very carefully in modifying or retouching his law of gravitation." But the debate was over for the rest of the attendees, and Sir J. J. Thomson, president of the Royal Society, and in this title Newton's successor, garnered general approval by concluding:

> This is the most important result obtained in connection with the theory of gravitation since Newton's day, and it is fitting that it should be announced at a meeting of the Society so closely connected with him [...] It is the result of one of the highest achievements of human thought.

He nevertheless regretted that this "highest achievement of human thought" was also one of the most incomprehensible, for "no one can understand the new law of gravitation without a thorough knowledge of the theory of invariants and of the calculus of variations."

As the meeting ended and the audience left the hall, commenting on the new results, there was much discussion of the rumor which had begun to circulate, according to which only three people in the entire world truly understood Einstein's general theory of relativity. As he was leaving, Silberstein, who had pleaded for caution but who was thought to be one of the rare experts on the theory of relativity, approached Eddington, who knew Einstein's theory in depth and who had initiated the observations of the 1919 eclipse. Silberstein congratulated Eddington on the success of the meeting, and on the fact that he was certainly one of the three people who understood general relativity. Eddington remained silent. "Don't be so modest, Eddington!" continued Silberstein, who was waiting for Eddington to return the compliment. "Not at all," replied Eddington, "I'm just wondering who the third one might be."

Suddenly Famous

The meeting of November 6, 1919 set off an avalanche of articles in newspapers the whole world over, and brought Einstein a sudden media-fueled glory, which would last until the end of his life. The day after the joint meeting of the Royal Society and the Royal Astronomical Society, the *London Times* carried the headline: "Revolution in Science—New Theory of the Universe—Newtonian Ideas Overthrown." Two days later, the *New York Times* gave an account of the London meeting and quoted, with some embellishment, Sir J. J. Thomson's conclusion: "one of the most momentous, if not the most momentous, pronouncements of human thought." The next day, the *New York Times* published a long article whose headlines had been chosen to intrigue the reader. For example: "Lights All Askew in the Heavens," "Men of Science More or Less Agog over Results of Eclipse Observations," "Einstein Theory Triumphs," and (to paraphrase) "A Theory which No More than 12 Men in All the World Could Comprehend."

Thus, the legend that only a few people in the world could comprehend Einstein's theory began to spread through the media like a wildfire. Let us take this opportunity to make two remarks. First, we note that, today, every scientific graduate student is capable of learning in a few hours the mathematical formalism of Einstein's general theory of relativity. Technically, it is a far less complex theory than quantum field theory or string theory. Nevertheless, this theory remains relatively difficult to access at the conceptual level. Every year sees the publication of a certain number of scientific articles which exhibit basic errors in the understanding of the theory. Moreover, some of the problems on the frontier of this theory (for example, those concerning the study of the interaction between two black holes and their gravitational radiation) pose formidable physical and mathematical difficulties, that only a small group of people in the entire world attack with success. The second remark we would like to make is that the particular mixture of conceptual subtleness with sophisticated mathematical formalism leads to the fact that today (perhaps more than ever), it remains difficult to popularize this theory, as soon as one wants to go any further than certain simplistic images or certain approximations which emasculate the very essence of the theory.

However that may be, the unique collection of circumstances and facts assembled in 1919 around Einstein's theory assured a durable (and justified) international celebrity for this theory and for its author. Among the circumstances leading to Einstein's celebrity we find: a major revolution in the fundamental concepts of reality (space, time, force, matter), a theory so deep and so new that even a majority of scientists confessed to not understanding it, a photogenic author with a sense of humor, a spectacular confirmation by the English of a theory developed in Germany. Finally, it was an opportunity to lift one's head towards the stars, after a terrible war in which so many innocents had died in the mud of the trenches.

"The Happiest Thought of My Life"

Let's return to that point of departure from which Einstein realized the necessity of generalizing the theory of relativity (from now on qualified by the word "special") which he had founded with his article from June 1905. Two years later, the theory of special relativity had attracted the interest of a number of prestigious (or destined to become so) scientists. The great experimental physicist Johannes Stark asked Einstein to write an article which would precisely map out his theory by explaining its principles, its consequences, and its contacts with experiment. It is in this article that Einstein commented on the experimental results of Kaufmann in the fashion that we have previously quoted. Einstein spent about two months on this review article. He was still making a living by working at the Bern patent office and thus only had a little time to dedicate to this task. Nevertheless, he profited from the spare moments in his workday to think about physics. It was while thus deeply reflecting on the significance of the principle of relativity that he had, one day in November 1907 at the Patent Office, what he called "the happiest thought of my life":

> I was sitting on a chair in my patent office in Bern. Suddenly a thought struck me: if a man falls freely, he would not feel his weight. I was taken aback. This simple thought experiment made a deep impression on me. It was what led me to the theory of gravity.

Let us explain the physical background of this revelatory thought. For this, we must go back to 1638, the date at which Galileo wrote his most important scientific work, *Discourses and Mathematical Demonstrations Concerning Two New Sciences*. By a remarkable combination of logical reasoning, thought experiments, and real experiments made on inclined planes,[3] Galileo was the first to think of the principle that one would today call "the universality of free-fall," or the "weak principle of equivalence." Let us cite the conclusion that Galileo reaches after a chain of reasoning where he (mentally) varies the relation between the density of the free-falling bodies under consideration and the resistance of the medium through which they fall: "Having observed this I came to the conclusion that in a medium totally devoid of resistance all bodies would fall with the same speed."[4] We note that this effect has indeed been directly verified by the first astronauts to set foot on the Moon, in July 1969. Profiting from the absence of atmosphere on the Moon (and thus of the resistance caused by the presence of a continuous medium), they dropped a hammer and a feather, and observed that these two objects fell at exactly the same rate.

Of course, physicists had not waited until 1969 to verify, with great experimental precision, Galileo's suggestion that in the absence of a resisting medium all bodies fall in the same way (with the same acceleration) in an external gravitational field. The first precise experiments are due to the great Newton himself, who compared the oscillations of two pendulums of identical exterior form, but of different composition and weight. Newton was also the first to understand that this property of the universality of free-fall was telling us something remarkable about gravitation. Indeed the second law of dynamics, introduced by Newton in 1686, says that a force F exerted on a body of mass m imparts to it an acceleration a, given by the simple formula $F = ma$. This formula tells us that a given exterior force F will not succeed in imparting the same acceleration to all bodies. For example, if a body A has a mass two times greater than a body B, the force F will impart an acceleration two times smaller to body *A* than to body *B*. One could thus say that the body *A* is two times more inert than body *B*. To sum up, Newton's second law of dynamics shows that the *mass m* (conceived by Newton as the quantity of matter) of a body measures the *inertia* of this body, that is to say its capacity to resist a change in motion.

We see also that, in general, the acceleration imparted by an external interaction has no universal qualities. For example, an electric field will communicate different accelerations to different objects. The acceleration of each object depends both on the value of its mass and on the value of its electric charge. In the same way, the acceleration communicated by a magnetic field has no universal qualities. From this point of view, it is remarkable that a gravitational field, such as the field describing terrestrial (or lunar) weight, imparts the same acceleration to all bodies placed at the same point in space. In the case of the gravitational field, the force it imparts to a body is called its *weight*. Newton thus understood that, among all the forces, the weight alone had the property of being exactly proportional to the mass. In other words, the force of *gravitation* is proportional to the inertia of the bodies on which it acts.

This deep and mysterious connection between gravitation and inertia was incorporated in a mathematical way by Newton in his theory of gravitation. Essentially, he declared that the mass played three separate roles: it measures the inertia of a body, it measures the fashion in which a body responds to an external gravitational field, and finally, it measures the fashion in which a body produces a gravitational field. For more than two centuries, following Newton's work, scientists ceased to be astonished by the remarkable fact that the mass thus played many *a priori* distinct roles.

Einstein's Elevator

However, on this November day of 1907, Einstein suddenly understood that this relation between inertia and gravitation must have a hidden significance, which should be illuminated. He thus began a chain of reasoning which would last eight years, taking him through many detours where he often got lost. In November 1915 he finally succeeded, with the foundation of a new theory of space, time, and gravity.

The first stage of this long journey towards the light was an innovative thought experiment. Generalizing his "muscular" intuition that a man in free fall no longer feels his weight, he imagined what one might observe in a *freely falling elevator*. Because of the universality of free-fall, any objects inside the elevator will "fall" with the same acceleration due to an external gravitational field. In particular, they will fall with the same acceleration as the elevator itself. Thus, with respect to the walls of the elevator, all these objects will have zero relative acceleration. In other words, they shall simply float freely without acceleration, either remaining forever at rest (if they are released with zero initial speed with respect to the interior of the elevator) or moving in a straight line at constant speed (if one has initially given them a certain speed with respect to the elevator's interior). This behavior has been rendered familiar to us through images from the exploration of outer space. This is what is referred to as the *antigravity* which reigns inside a space capsule in free fall in the Earth's gravitational field. In physical terms, we say that the interior of an elevator or a space capsule defines a *system of reference*, or more simply, *a reference frame*. An elevator in free fall thus defines a *reference frame in free fall*. One may thus summarize the observations that one can make from the interior of such a reference frame by saying that the exterior gravitational field is *effaced* within a freely falling reference frame.

Einstein also considered another situation: that in which there is no external gravitational field. Let us say, for example, that instead of moving near the Earth's surface (where the mass of the Earth creates a gravitational field) one moves somewhere very far from any mass: far from the Earth, far from the Sun, far from our galaxy and from all other galaxies. (This is a thought experiment, where one can imagine the existence of such a region.) Let us consider once again an elevator located in such a region where there is no "real" gravitational field. Einstein then imagines that one accelerates this elevator by pulling it with a force directed in a certain direction, that we shall call "up." Inside this elevator, accelerating "upwards," a process will occur which has been rendered familiar to us by various means of transportation: for example, in a car accelerating forwards, the passengers are shoved back into their seats, and any unattached objects move, with respect to the vehicle's interior, by accelerating towards the back. Thus, inside an "upward" accelerating elevator, all objects will be accelerated "downwards." This acceleration is universal: the same for all bodies, whatever their mass and composition may be. Thus this acceleration exactly resembles the universal acceleration which would be imparted by a "true" gravitational field outside the elevator.

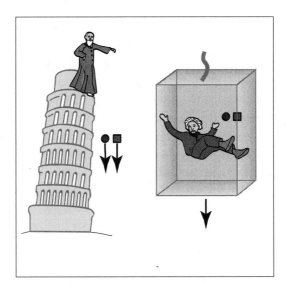

Figure 6. Universality of free-fall (Galileo) and the principle of equivalence (Einstein).

Einstein concludes from this that everything occurs as if the acceleration imparted to the elevator had, viewed from the interior of this elevator, created an apparent gravitational field. These thought experiments showed him the intimate connections between gravitation and inertia: by using various effects of acceleration (and thus the *inertial* properties of bodies), one could either efface a real gravitational field, or create an apparent gravitational field.

Towards a Generalized Relativity

Why would Einstein consider the various possible effects arising from the interplay between inertia and gravity to be important, when he was writing a review article on the 1905 theory of special relativity? Let us recall the central intuition of the principle of relativity, such as Galileo had formulated it: "the motion is as nothing." However, this statement only refers to motion in a straight line and at constant speed. Everyone knew that it is only in a ship which moves in an unchanging direction on smooth water that one does not detect the effects of the overall motion. When the ship turns brusquely, or accelerates, the passengers in the cabin will feel the effects. Thus, before Einstein, everyone thought that the principle of relativity could only be applied to linear and uniform relative motion. However, Einstein understood that he could generalize the principle of relativity to *accelerated* motion (in either a straight line, or following a curve, as in a bend in the road). However, to be able to envisage such a generalization, it was necessary to take gravitation into account. He could not say, "an accelerated motion is as nothing," but he could say that an accelerated motion is like a gravitational field. In other words, there is an *equivalence* between acceleration and gravitation.

As we have said, Einstein's scientific methodology consisted, when possible, of assuming as a departure point some general principles which would permit one to constrain the laws of physics. Thus, in 1907 he proposed a new physical principle: the *principle of equivalence* between gravitation and acceleration (or between gravitation and inertia, since the apparent effects of an external acceleration are called inertial forces). In Einstein's hands, this principle turned out to be an extraordinary tool for constructing, between 1907 and 1915, a generalization of the 1905 theory of relativity. This theory is called the generalized theory of relativity or, more simply, the theory of general relativity.

Einstein's Theory in a Phrase and an Image

One can summarize the theory of general relativity, or Einstein's theory of gravity, in one phrase: *space-time is an elastic structure which is deformed by the presence, in its midst, of mass-energy.*

We shall try to help the reader understand the meaning of this phrase step by step, without making use of equations or other mathematical formulas. It will perhaps be useful to suggest to the reader, from the very beginning, a certain image of an elastic structure deformed by the presence, in its midst, of matter. This image shall, of course, be incomplete, and deceitful in certain aspects, but we shall try to render it as faithfully as possible to the image of space-time given by Einstein's theory.

The image that we wish to suggest to the reader is not the simplistic image which appears in many articles and books of popularization: that of a massive ball placed on a rubber sheet, and deforming it with its weight. This image indeed contains certain aspects which are analogous to what happens in Einstein's theory, but it has the grave problem of also containing some very deceitful features. For example, it suggests that the deformation of the sheet can only be thought of as a curvature in a space exterior to the sheet, and also that this deformation only exists thanks to the external gravitational field acting on the ball. On the other hand, that which indeed gives Einstein's theory its flavor is that the deformation of space-time is an affair purely intrinsic to this space-time, and has no need of extra dimensions in order to be formulated.

Note also that we will try, as much as is possible, to avoid the use of the word "curvature" of space-time, or the expression "curved space-time." Indeed, for the majority of people, the word "curved" immediately evokes the image of a line or a surface which is curved within a larger, external space, as in the surface of a sphere in ordinary (three-dimensional) Euclidean space. On the other hand, the *curvature* spoken of by Einstein's theory is not a curvature of this type (even if a sphere is still curved in the sense required by Einstein), but rather an *internal deformation* which has no need of external dimensions in order to exist. This is why we shall replace everywhere the word "curvature" by the word "deformation" and the adjective "curved" by "deformed." We hope to thus avoid having the reader's mind blocked up by misleading images.

The image that we suggest, as an analog of the space-time structure of Einstein's general relativity, is a culinary image: *veal in aspic!* More precisely, a block of aspic (gelatin) which contains in its midst long shreds of veal, as well as perhaps other things (such as pieces of vegetables). The gelatin symbolizes space-time. The reader can well imagine that if this gelatin contained neither meat nor vegetables, it would have a homogeneous and isotropic structure, that is to say it would have the same properties everywhere, and in every direction. This *uniform* state of the aspic is the analog of the chronogeometric structure of Minkowski space-time, such as it is represented in Figure 7, in the form of a regular and uniform network of spatio-temporal hourglasses. We shall consider this state as the nondeformed state of the gelatin (and of space-time).

We may then consider many distinct ways of deforming this uniform state. If one shakes, in an oscillatory manner, one of the edges of the block of gelatin, the gelatin vibrates and begins to quiver and, in fact, to be crossed by vibrational waves. These waves of elastic vibration within the gelatin will have a direct analog in the case of space-time, which allows for the propagation of waves of deformation of its structure, called gravitational waves. One could also be contented with compressing the aspic in a stationary way by squeezing opposite sides of the block. This will of course deform the interior of the block of gelatin, in an anisotropic fashion: certain directions being compressed, and others stretched. Finally, the long fibers of meat which traverse the block of aspic are analogs to the world-lines traced by material particles within space-time (see Figure 7). One can imagine that the gelatin in the

Figure 7. Chronogeometric structure of the Poincaré-Minkowski space-time (repeated from Figure 3).

immediate vicinity of the veal fibers is denser or simply more full of nutrients than plain gelatin, without meat. This will be the analog of the fact that space-time is more deformed the closer it is to a distribution of mass-energy.

Space-Time Deformed

Let us return to the theory of general relativity and define the notion of deformed space-time. We first recall the chronogeometric structure of a nondeformed space-time: that of the theory of special relativity, such as was introduced by Poincaré and Minkowski. This structure is given through the notion of the squared interval between two space-time points, that is to say two events. The squared interval between any two points is obtained, through a generalization of the Pythagorean theorem, as the algebraic sum of four squares: three of these squares (those of the differences in length, width, and height between the two events) appear in this sum with a plus sign, while the fourth square (that of the difference in date, multiplied by the speed of light) is given a minus sign. From this, the ensemble of points separated from a given point by a squared interval equal to +1 form, in space-time, not a (hyper-)sphere, but a (hyper-)surface which resembles an hourglass, that is, two cones joined by a throat.[5] The structure of such a nondeformed space-time is homogeneous, which is to say that it is the same everywhere in space-time. Whichever event we may pick, around which we locally examine the space-time, we will see the same structure. Moreover, this structure is isotropic, in the sense that there is no direction within space-time which plays a special role. Here, the reader who looks at the checkerboard of hourglasses representing this structure (Figure 7) will perhaps say to herself that it seems to have, at each space-time point, a privileged direction. Indeed, each hourglass appears to have an axis of symmetry: the vertical axis, passing through the center of the hourglass, around which one may rotate the hourglass without this rotation affecting its visual appearance. However, in reality, this apparently privileged axis is an artifact of the visualization of space-time within a three-dimensional space which human sight intuitively interprets as a Euclidean space. Indeed, the vertical axis in this visualization represents the world-line of an observer who is "at rest" in space, but who has a continuous existence in time. On the other hand the principle of relativity tells us that such an observer does not define a privileged reference frame. Any other observers moving at constant speed with respect to this particular observer will see an identical space-time structure. The world-lines of these observers "in motion" (but moving more slowly than light) will be straight lines inclined with respect to the "vertical" at an angle less than $45°$. As a consequence, if one considers a particular hourglass, all of the straight lines passing through its center and remaining "inside" the hourglass (that is, without piercing its interior surface) will be axes of symmetry for it, and none of them play a privileged role. Such is the homogeneous and isotropic structure of nondeformed space-time gelatin.

What is the chronogeometric structure of a deformed space-time (which one generally refers to as being curved)? It is a structure where the duration-distance between two events is still defined by a certain squared interval, but where, contrarily to the case of Minkowski space-time, this squared interval is given by a very complicated mathematical formula when the two events are far from each other. On the other hand, if one considers events that are very close together (both in space and in time), the squared interval is given by a simple enough mathematical formula, even if it is still more complicated than the one relevant for Minkowski space-time. As Einstein understood it in 1912, the squared interval between two events in a deformed space-time is quite analogous to the distance squared between two points on a curved surface embedded in ordinary Euclidean space.

As an example of a curved surface, let us take the surface of the Earth. If you consider a small portion of the terrestrial surface, for example, a surface of one meter squared, you could generally identify it with a small portion of a plane (it would suffice to consider the plane tangent to a point situated near the center of the small surface portion considered). Therefore, the distance squared (the square of the distance) between two points on this small surface will be, to a very good approximation, equal to the distance squared between two points in a plane. This can be obtained by using the Pythagorean theorem. The only complication comes from the fact that it is no longer possible to crosshatch the entire surface of the Earth, with its valleys and its mountains, by a perfectly regular network of coordinates (such as length and width).

On a planar surface, for example, a sheet of paper placed on a table, it is easy to pinpoint the location of points on the surface by using a regular orthogonal grid, such as one uses for graph paper. Such a regular grid becomes impossible to realize on a surface having bumps and hollows. To fix an arbitrary point, on any curved surface whatsoever, we thus use two numbers, let's say x and y, which will no longer have the simple meaning of a length and a width. For example, on the surface of the Earth, one could use the longitude as the first coordinate x, and the latitude as the second coordinate y. Note that one may use such coordinates even when one cannot approximate the Earth's surface as a sphere, for example, on a mountain, or at the bottom of a canyon. There is no need to give a third coordinate (let's say the altitude measured from sea-level), since the first two coordinates (longitude and latitude) suffice to fix one's location on Earth, and the altitude is a determined function of the longitude and latitude. However, a moment's reflection will show that if one considers the grid defined by the longitude and latitude on a small part of the Earth's surface, taken on the side of a mountain, or on the slope leading into a canyon, this grid will be a deformation of the regular grid on graph paper: one will always have a tiling of the surface by two families of lines, but each tile, in place of being square, will be a sort of parallelogram; its sides will not all be equal, and will not meet at right angles.

Locally, as we have said, one can associate a fragment of this tiling to a regular tiling of parallelograms drawn on a plane tangent to the surface. The Pythagorean theorem, generalized to triangles which are no longer necessarily right-triangles, then tells us that the distance squared between two joints in this (planar) tiling is given by a sum of squares and double products of the coordinate differences between the two joints. To define the distance squared between neighboring points of any curved surface whatsoever, whose points

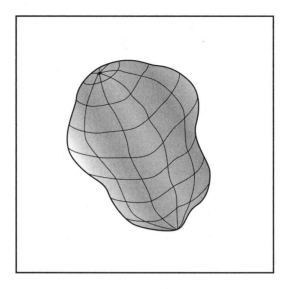

Figure 8. Grid defined on a curved surface by two coordinates x and y.

are fixed by two coordinates x and y, one must thus give, at each point, three quantities: the coefficient in front of the square dx^2 of the difference dx between the first coordinates of the two points, the coefficient in front of the square dy^2 of the difference dy between the second coordinates, and the coefficient in front of the double product $2dxdy$. (We consider the mathematical limit where the points are infinitely close, from whence the notation d which designates an infinitesimal distance.) These three coefficients determine the geometry of the surface considered, and for this reason are denoted, respectively, by g_{xx}, g_{yy}, and g_{xy}, where the letter g recalls the first letter of the word "geometry."

During his studies at the Zurich Polytechnic, Einstein had very much appreciated Carl Friedrich Geiser's course on the infinitesimal geometry of surfaces. Geiser had lectured on the theory developed by the famous mathematician Karl Friedrich Gauss, the theory which we have just explained, where one considers the squared distance between points which are infinitely close together. In 1912 he thus remembered that the geometry of a deformed (or nonplanar) surface may be defined by giving the three quantities g_{xx}, g_{yy}, g_{xy} at each point on the surface. The geometric tensor, or more simply the metric tensor, **g**, is the structure defined by giving, at each point on the surface, the three quantities g_{xx}, g_{yy}, g_{xy}. Einstein realized that he needed a generalization

Figure 9. Chronogeometric structure of a deformed space-time.

of this notion to the case where the (two-dimensional) surface was replaced by a (four-dimensional) space-time. Bernhard Riemann, Gauss' student, had already generalized Gauss' theory to deformed spaces of any dimension whatsoever. But Riemann had only considered the case of spaces which resembled ordinary Euclidean space in the neighborhood of each point. In other words, of spaces in which the locus of points separated from a given central point by a small distance (squared) ε^2 had the form of a deformed (hyper-)sphere, which is to say a sort of (American) football.[6] Einstein understood that he must generalize Riemann's theory to the case where the locus of points separated from a given central point by a small (positive) squared interval ε^2 had the form of a deformed hourglass.[7]

A deformed space-time is thus defined by giving, around each point, such a deformed hourglass. See the representation of this concept in Figure 9, and compare it with the nondeformed case of the space-time of special relativity (Figure 7). Einstein subsequently understood that such a deformed space-time could not be tiled analogously to the regular grid of graph paper, through the use of the four regular coordinates (length, width, height, and date) which he had used in the special theory of relativity. As in the case of the Earth's surface, one had to use more general coordinates (analogous to the longitude and latitude on a sphere having bumps and hollows; see Figure 8). As space-time is four-dimensional, one needs four coordinates to pinpoint a given event.

One may designate these coordinates in various ways, for example (x, y, z, t), (x_1, x_2, x_3, x_4), or (x_0, x_1, x_2, x_3).

Einstein understood (after several years of confusion, hesitation, and doubt) that there was total mathematical freedom in the choice of these four coordinates, in other words that no system for fixing points in space-time was favored *a priori*. He thus ended up postulating that the laws of physics should take the same form in whichever system of coordinates may be used. He called this postulate the *principle of general relativity* since he initially thought that this postulate was a generalization of the 1905 principle of relativity, which was restricted to consideration of the coordinate systems used by observers in uniform motion with respect to each other.[8] The imposition of this postulate permitted him to quite strongly constrain the possible form of the laws of relativistic gravitation. It was thus that he was able to make his most beautiful discovery, that which J. J. Thomson, Dirac, and many other physicists considered as the greatest accomplishment in the history of human thought: the theory of general relativity, or Einstein's theory of gravity.

To summarize, the first stage of the theory of general relativity consists of saying that the chronogeometry of a deformed space-time is given by the structure represented in Figure 9: the collection of events separated from a given central event by a (positive) infinitesimal squared interval ε^2 is a deformed hourglass (or, in mathematical terms, a general hyperboloid). To precisely describe this structure, one must provide, at each space-time point, a mathematical object, called **g**, which is called the chronogeometric tensor or, more simply, the metric tensor. This tensor is a collection of ten coefficients which determine the form taken by the Pythagoras-Einstein theorem in an arbitrary system of coordinates.[9] We note that, by a happy coincidence, the notation **g** will call to mind the *geometry* of space-time and *gravitation* all at once.

Einstein's Law of Space-Time Elasticity

To help one understand the intuitive meaning of Einstein's theory of gravity, let us recall the theory of elasticity, due to the British scientist Robert Hooke. Hooke was one of the most fecund scientific minds of the seventeenth century. He made seminal contributions in an impressive number of different scientific domains, and was for a long time the secretary of the Royal Society. His work anticipated many of Newton's discoveries, both on the general laws of dynamics and on the $1/r^2$ behavior of the law of gravitation. Unfortunately for him, Newton, who was a genius but also easily offended and irascible, turned against him and did all that he could to minimize the importance of his work. Newton would have certainly been furious with the presentation we shall make of Einstein's theory of gravity (which supplanted his own) in the form of a generalization of Hooke's law of elasticity!

The point of departure for Hooke's theory is quite simple to understand. Let us consider an arbitrary elastic structure, which by definition has the tendency to return to its initial form, after being deformed by a force which acts upon it. A simple example of an elastic structure is a spring. We consider a spring, whose top is fixed to a rigid, massive body, and whose bottom is free. We pull downwards on, or attach a weight to, the bottom end of the spring. The spring will deform and stretch out. If one attaches a reasonable (not too heavy) weight, one notices that the amount that the spring stretches is directly proportional to the weight which is suspended: twice the weight will produce twice the stretch. In other words, the *deformation* of an elastic structure is proportional to the *tension* which acts on this structure. If we denote the deformation by the letter D and the tension by the letter T, Hooke's general law of elasticity is simply $D = \varkappa T$, where $\varkappa$ is a coefficient of proportionality which measures the *elasticity* of the structure considered. The larger $\varkappa$ is, the more elastic the structure is, which is to say the more it is deformed under the influence of a given tension. One could also say that the inverse of the coefficient $\varkappa$, the quantity $1/\varkappa$, measures the *rigidity* of the structure considered. The smaller $\varkappa$ is, the larger the rigidity (and the smaller the elasticity). This general law of elasticity only holds in a limited range (centered around zero) of applied tension. Note that the tensions, and the associated deformations, can go in one direction or the other, that is, they can be positive or negative. Whatever the sign of the tension applied may be, the deformation will return to zero when the tension is returned slowly to zero. It is the primary characteristic of an elastic structure that it returns to its initial nondeformed state when the force which deforms it ceases to act.

On the other hand, when one passes a certain threshold (called the limit of elasticity), specifically if one applies tensions that are too strong, one will in general leave the region of elasticity of the structure. One then enters the region of *plasticity*, where the structure acquires a permanent deformation, which remains when the tension ceases acting, and then finally the region of *rupture*, where the structure breaks.

To strengthen our intuition, and to get us closer to our model of space-time gelatin, let us consider an elastic structure which is a three-dimensional medium, for example, the medium provided by veal in aspic. That which we are about to say will apply as well to the case of a solid medium like a metal, but the rigidity of metal is so high that one's intuition has a hard time considering a metallic block as an elastic structure. We therefore consider a block of (homogeneous) gelatin. We deform this block by applying pressure or tension along its external faces. This will create a state of tension, or state of stress within the block. This state of stress is measured (in the mechanics of continuous media) by a mathematical object called the *stress tensor*. This tensor, which we shall denote by $\mathbf{T}$ for tension,[10] permits one to calculate the forces which act on the surface of a hypothetical volume cut out from the interior of the medium, as exerted by the exterior of the volume considered. For a gaseous medium, $\mathbf{T}$ is determined by the pressure which is present within the gas.

It remains for us to describe how one defines the deformation $\mathbf{D}$ of a block of gelatin. The deformation $\mathbf{D}$ is defined, when it is small, as the difference between the geometric structure of the deformed block and that of the initial nondeformed block. How does one measure the geometric structure of a continuous medium? In precisely the same way that we have above visualized the geometric structure of a space. We first visualize the geometry of the nondeformed block of gelatin (considered in the usual Euclidean space) by representing, around each point in the block, the locus of points situated at unit distance from this point. This gives us a regular network of spheres, within the block's interior. Now, we deform the block, that is we make the gelatin move (continuously) in an arbitrary way (much as the contents of a toothpaste tube are deformed when one presses on the sides to squeeze out the toothpaste). This continuous deforming motion of the gelatin will deform the network of spheres. First, the center of each sphere will move. However, such an effect is not in itself connected to the tensions within the medium, since, for example, one could move the entire block of gelatin one centimeter to the right through an overall translation, without creating any stress in the medium. The important thing, from the point of view of elasticity, is thus to measure how each sphere is deformed when it follows the motion of the gelatin surrounding it. If one considers, as we do here, small displacements, one finds that a sphere is deformed into an ellipsoid, which is to say a sort of football or squashed sphere. We shall thus call the deformation, $\mathbf{D}$, the mathematical object which measures the difference between an ellipsoid and a sphere. One may show that this object has the same mathematical structure as the object which measures the presence of stresses within the medium, and is thus a tensor, called the deformation tensor.[11] Finally, the law of elasticity for a homogeneous and isotropic continuous medium, such as a block of pure gelatin, is obtained by writing the most general linear relation[12] $\mathbf{D} = \hat{\varkappa}\mathbf{T}$, which can exist between two mathematical objects of the same type: the deformation tensor $\mathbf{D}$ and the stress tensor $\mathbf{T}$.

Having improved our understanding of the elasticity of a continuous medium (in the sense of ordinary mechanics), we can now return to the central aim of this chapter: to understand general relativity as a theory of space-time elasticity. For this, we must discuss two things: (i) what is the analog of $\mathbf{D}$; what mathematical object measures the deformation of a space-time with respect to the uniform space-time of Minkowski; and (ii) what is the analog of $\mathbf{T}$, in other words what is the mathematical object which measures the *cause* (or *source*) of the space-time deformation; that element without which space-time would remain a Minkowski space-time. The answer to (ii) was obtained fairly rapidly by Einstein through the following reasoning.

First, Einstein suggested identifying the metric tensor $\mathbf{g}$, which measures space-time chronogeometry, with the gravitational field. This follows from the study of the principle of equivalence, discovered by Einstein in November 1907, and discussed above. Consider, for example, the simple case of Minkowski space-time. If an observer examines this Minkowski space-time while remaining in an inertial, nonaccelerating, reference frame, he will observe no gravitational field (free particles will not "fall," but remain at rest or move at constant speed), and the metric tensor $\mathbf{g}$ which measures the space-time chronogeometry will be trivially simple, and given by constant coefficients.[13] On the other hand, an observer placed within an accelerating elevator, using coordinates related in a nonlinear fashion to the usual coordinates of special relativity, will observe two correlated phenomena: (i) the metric tensor will be more complicated than before, with coefficients $\mathbf{g}$ which vary from one point to the other, and (ii) there is an apparent gravitational field in the accelerating elevator: particles will fall with a nonzero acceleration. The acceleration caused by the apparent weight is directly connected to the fact that the coefficients $\mathbf{g}$ vary from one point to the other.

Having understood that $\mathbf{g}$ = chronogeometry = gravity, the second step consists in understanding what is the source of $\mathbf{g}$. This must be the source of gravitation. However, ever since Newton's work, one knows (because of the universality of free fall and the equality between action and reaction) that the mass measures both that which gravity acts on (determining the weight), and that which creates a gravitational field. Thus the source of the gravitational field is, for Newton, the mass. However, as was discussed in the second chapter, the theory of special relativity has completely modified, and enriched, this concept of mass. It is replaced by the concept of mass-energy, which is what remains conserved through all those transformations in which, by virtue of the relation $E = mc^2$, some mass can be transformed into energy, or vice versa. Thus, Einstein expected that the source of gravitation would be the presence within space-time of a mass-energy distribution. Our search for the source of gravitation cannot, however, stop there. A more detailed study of the origin of the conservation of mass-energy, using the special theory of relativity, shows that the density per unit volume of mass-energy is only one component of a more complicated mathematical object, called the stress-energy-momentum tensor, often called simply the stress-energy tensor. This tensor has ten components: one component measures the mass-energy density per unit volume, three others measure the momentum (or quantity of motion) density per unit volume, and the remaining six measure the stress tensor, in precisely the same way as the previously introduced stress tensor measures the stresses in the interior of three-dimensional continuous media. We shall continue to call this ten-component tensor[14] **T**, which simultaneously generalizes both the mass density (mentioned in Newton's law) and the stress tensor (mentioned in Hooke's law).

Let us return to one of the crucial moments in Einstein's invention of general relativity. As we have said, Einstein came up with the initial idea of a generalized relativity in 1907, while he was still working (eight hours a day, including Saturdays) at the Bern Patent Office. Soon, however, the great interest aroused by the theory of special relativity of June 1905, and by some of his other works, led several scientific centers to offer him a university post. In 1909, he resigned from the Bern Patent Office to accept a post as an associate professor at the University of Zurich (with the same salary which he had in Bern). Einstein and Mileva were happy to return to Zurich, the town where they had met while both studying at the Polytechnic. Their second son, Eduard, was born there in 1910. Nevertheless, in 1911, Einstein accepted a post, this time as a full professor, at the German university of Prague. He would only stay one year in Prague. There he frequented the literary salon of Bertha Fanta, where he met the (Jewish) writers and thinkers of Prague, notably Max Brod and Franz Kafka. It was in Prague that he renewed his search for a generalization of relativity (during the years 1907–1911, he had dedicated himself primarily to the development of his ideas on quanta; see below). There he obtained some quite important results, notably a more precise understanding of his *principle of equivalence* and the idea that this principle implied an observable deflection of light rays passing near the edge of the sun,[15] and a shift towards the red, that is to say toward lower frequencies, of light rays emitted from the surface of a massive body, such as the Sun.

At the end of July 1912, Einstein and his family returned to Zurich. Einstein had accepted a position as a full professor at his former school: the Polytechnic, from now on promoted to the rank of Swiss Federal Institute of Technology, or Eidgenössische Technische Hochschule (abridged as ETH). It was in Zurich, it seems, in the month of August 1912 that Einstein made a very important conceptual step towards the construction of general relativity. He essentially understood what we have explained above, that (i) the gravitational field is equivalent to a deformation of the space-time geometry, and must therefore be described by the ten components of the chronogeometric tensor g; that (ii) the source of the "g field" is the distribution of mass-energy, momentum, and stress, described by an object with ten components, the stress-energy tensor T; and finally that (iii) the fundamental equation of relativistic gravitation must have the form of a law of space-time elasticity:[16] $D(g) = \varkappa T$, where $D(g)$ is a mathematical object constructed from g, which is presumed to measure the deformation of space-time, in other words how a space-time having a geometry described by g differs from Minkowski space-time.

With this program in mind, he went to see his old friend Marcel Grossmann, who had been his comrade on the benches of the ETH (the ex-Polytechnic), who had "saved his life" by providing his notes before the exams, had later helped him to find work at the Bern Patent Office, and who had worked hard to get the ETH to offer him a professor's chair. Marcel Grossmann was a mathematician who had become, in 1907, a full professor of geometry at the ETH, and since 1911 Dean of the ETH college of mathematics and physics. Einstein proposed that Grossmann collaborate with him, to find a "good definition" of the mathematical object $D(g)$. Grossmann taught geometry at the ETH, and his mathematical work was also concentrated on the problems of geometry, but it was geometry in the sense of studying the structures defined by collections of straight lines and points in homogeneous spaces. Grossmann was not familiar with the type of nonhomogeneous geometry which Einstein had need of. He made some searches in the mathematical literature, and rapidly found that several works, due to the mathematicians Riemann, Christoffel, Ricci, and Levi-Civita, would doubtless contain the mathematical tools necessary and sufficient to construct the object $\mathbf{D}(\mathbf{g})$ searched for by Einstein. However, these mathematical tools were of great complexity and, to master them and understand their physical implications, Einstein and his friend had to apply an intense effort for several months, and even (for Einstein) years. We quote an extract of a letter from Einstein to his colleague Arnold Sommerfeld, written during the period when Einstein was deploying "frankly superhuman" effort, in his terms, to the problem of a relativistic theory of gravitation:

> I am now working exclusively on the problem of gravitation and I think that, with the help of a mathematician friend that I have here, I will be able to overcome all the difficulties. However, one thing is sure—never before have I ever been as tormented by a problem. I have acquired a great respect for mathematics, since before I had the tendency to consider sophisticated mathematical techniques as a useless luxury! In comparison with the present problem, the theory of special relativity was child's-play.

In fact, Einstein met some unexpected technical difficulties, which stopped his collaboration with Grossmann from completely resolving the difficulties and constructing the object $\mathbf{D}(\mathbf{g})$ which they sought. They came very close to their goal, and considered for a moment a candidate to describe $\mathbf{D}(\mathbf{g})$ which was essentially the right one,[17] which Einstein then rejected because of an apparent conflict between the principle of general relativity which he had postulated, and the principle of causality. This obliged Einstein to devote three more years of "frankly superhuman" work before finding the definitive solution, in November 1915, in Berlin. In fact, Einstein had left Zurich and the ETH in 1914 to occupy a post as research director, without teaching duties, at the Prussian Academy of Sciences in Berlin. This post had been created especially for him, at the initiative, notably, of Max Planck, whom we have previously encountered as the first high-level physicist to understand that the theory of special relativity was a major conceptual revolution, comparable in size to the Copernican revolution.

To conclude this chapter, I ask for a final effort from the reader, who has bravely weathered the long discussion made above of the description of a warped space-time geometry, described by the ten components contained in $\mathbf{g}$, and of the search for an object $\mathbf{D}(\mathbf{g})$, also with ten components, measuring the deformation that $\mathbf{g}$ represents with respect to the nondeformed case of a Minkowski geometry. If the reader recalls that Einstein had to supply an intense and continuous effort of concentration during five consecutive years

(1911–1915), he or she will perhaps have the heart to concentrate for a few more minutes to acquire some intuition about one of the highest summits attained by human thought. I hope indeed that the reader of this book will put the lie to the passage written by Hannes Alfven on the occasion of the celebration of the centennial of Einstein's birth:[18]

> Many people probably felt relieved by being told that the true nature of the physical world could not be understood except by Einstein and a few other geniuses [...] Paradoxically enough, it is possible that the general public had acclaimed Einstein, not because he was a great thinker, but because he freed everybody from the task of having to think.

Riemann's mathematical research, and that of his successors Christoffel, Ricci, and Levi-Civita, on general spatial (or space-time) geometry had produced many mathematical objects measuring the difference between a deformed space (often called curved) and a rigid and uniform space (called flat). Visually the problem is to describe the difference between Figure 7 and Figure 9. Contrarily to what a naive approach (which sufficed in the simpler case of the deformation of a block of gelatin) might have led one to hope, it does not suffice to take the difference, at each point, between the metric tensor **g** of the deformed geometry and the simple value that this tensor takes in a nondeformed space (or space-time), consisting only of coefficients equal to +1 or −1. Indeed, the thought experiment of Einstein's elevator shows that if one uses an accelerating reference frame (or just as well, as Einstein also remarked, a rotating reference frame), the metric tensor describing the chronogeometry of Minkowski space-time in such a reference frame will take a quite complicated form, with coefficients **g** which vary from point to point.

We can now return to the initial formulation of the "happiest thought" of Einstein's life. We place ourselves in the midst of an arbitrary space-time, deformed by the presence of matter and of stresses, and supplied with a nontrivial metric tensor **g**, and we let an elevator fall freely within this space-time. The initial idea was that the gravitational field was *effaced* within the interior of such a freely falling elevator, which is to say erased within a reference-frame defined by the elevator's interior (see Figure 6). Is this effacement complete? No, since two objects situated within the elevator's interior will not fall with *exactly* the same acceleration. Indeed, not being situated at the same point of space(-time), they are subject to slightly different gravitational accelerations (which differ as well from the acceleration of the falling elevator itself, which depends on the position of the center of mass of the elevator). Thus, in the interior of a freely falling elevator, a small residue of the gravitational field continues to exist: that part which the elevator's fall does not succeed in effacing because of variations from point to point of the gravitational field.

If one was working within the framework of the Newtonian description of gravity, this uneffaced residue would be described by what are called *tidal forces*. This name is derived from the following fact: the Moon (as well as the Sun) gravitationally attracts the Earth (at the same time that the Earth attracts the Moon). Thus the Earth falls towards the Moon (and the Sun), and vice versa. The Earth thus forms, in a natural fashion, an elevator in free fall. The fall of Earth towards the Moon effaces the greater part of the acceleration produced by the Moon acting on Earth. Now consider a case where the oceans cover the entire Earth. That part of the ocean situated on the moonward side of Earth will be attracted towards the Moon more strongly than the center of mass of the Earth, which in turn will also be attracted more strongly than that part of the ocean situated on the side opposite the Moon. From this, there remains a differential effect, on the Earth in free-fall, which lifts the moonward side of the ocean towards the Moon, and which lifts the ocean on the opposite side of the Earth in an inverse sense, that is to say even farther from the Earth. This effect, which raises two bulges of ocean on opposite sides, is the cause of the tides (which are thus determined by the two gravitational field residues caused by the Moon and the Sun in the reference frame which falls with the Earth).

The mathematical theory (initiated by Riemann) of the incompletely effaced residue of a chrono-geo-gravitational field **g** in a reference frame in free-fall led to the appearance of a mathematical object having 20 components: the Riemann-Christoffel tensor, **R**. This generalization of the tidal tensor[19] of Newtonian gravitation is the most complete possible measure of the true local deformation of a curved space-time. But this could not be the mathematical object searched for by Einstein, which must have only ten components, like its source, **T**. After some hesitation and confusion, Einstein understood, in November 1915, that there was only one way of constructing an object from **R** with ten components measuring, in an incomplete fashion, a space-time deformation, while respecting both the principle of general relativity and the conservation of energy and momentum. This ten-component object, which we here denote **D(g)**, is called the Einstein tensor.[20] Thus, he could finally write, after eight years of research, Einstein's equations of gravitation, $\mathbf{D(g)} = \chi\mathbf{T}$, where the ten quantities on the left side of the equation measure (in a partial fashion) the locally measurable deformation of the space-

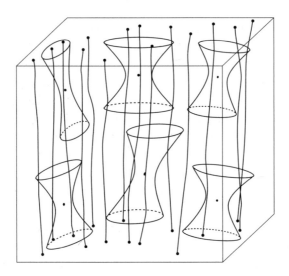

Figure 10. Space-time deformation caused by the presence of mass-energy and stresses.

time chronogeometry, while the ten quantities on the right contain the source of this deformation: the distribution of stresses, momenta, and mass-energy. As we have already said, these ten equations, connecting a deformation to the presence of stresses applied in the midst of the medium, are analogous to the fundamental equations governing the elasticity of a slightly deformed medium.

Figure 10 illustrates what Einstein's gravitational equations tell us. The presence of mass-energy is here visualized by drawing the world-lines (or space-time lines) that particles trace in space-time. Note the fibrous character of this distribution of mass-energy. The presence of this distribution produces a deformation of space-time geometry, shown by a network of deformed hourglasses.

The Strength of Space-Time Elasticity

We hope that at this stage the initial image which we proposed shall have acquired greater richness: space-time as gelatin, and the deformation-generating matter as fibers included in the midst of this gelatin. To finish, it remains for us to fix the value of the coefficient $\varkappa$, appearing in Einstein's equations, and measuring, as we have said, the elasticity of the space-time gelatin. Einstein was able to determine this coefficient by demanding that, in a certain approximation, the ten equations summarized by the expression $\mathbf{D(g)} = \varkappa\mathbf{T}$ reproduce the Newtonian theory of gravity, with a single gravitational potential from which follows a force inversely proportional to the square of the distance. He found that $\varkappa = 8\pi G/c^4$, where G is Newton's gravitational constant appearing in the gravitational force $F = Gmm'/r^2$ between two masses m and m' separated by a distance r.

When one uses the usual units of theoretical physicists, by measuring distances in centimeters, durations in seconds, and masses in grams, one finds that the numerical value of space-time elasticity is approximately two times ten to the power minus 48, that is, $\varkappa = 0.000, 000, 000, 000, 000, 000, 000,$ $000, 000, 000, 000, 000, 000, 000, 000, 002!$ This shows well that the elasticity of space-time is extremely small or, equivalently, that the rigidity of space-time, measured by the inverse $1/\varkappa$ of the elasticity, is extremely large. This explains why, for millennia, one was able to suppose that space and time were "rigid" structures, uninfluenced by the presence of energy or stresses. One must concentrate enormous densities of energy or stress to succeed in deforming the space-time gelatin in a significant way.

4

———

Einstein's World-Game

Why does he play, the great Child whom Heraclitus had seen within
cosmic Time ($\alpha\iota\omega\nu$), the Child who plays the Game of the world?
—Heidegger, *The Principle of Reason*

Mercury Advances, the Sphinx Speaks

Berlin, Germany, November 1915

The month of November 1915 marks the birth of general relativity, and thus
the birth of a new world, in the sense that Minkowski had used this word (*die
Welt*): the world of space-time. Einstein's new world is not, as had been believed for two thousand years, a rigid checkerboard on which matter and force
play, without their game having any influence on the checkerboard which
carries them. Instead, Einstein's new world takes an active part in the game
played by matter and force. The new game of the world is a game for four, that
is to say space, time, matter, and force, or rather a game for two, space-time
and mass-energy, where all the partners influence each other reciprocally. The
mass-energy deforms space-time by its presence, and the deformed checkerboard of space-time determines the fashion in which mass-energy meanders
through it.

When did Einstein, a modern Ulysses, know that he had finally arrived
at safe harbor after eight years of wandering, confusion, and confrontation
with various obstacles? One can date this moment with high precision. It was
between the 11th and the 18th of November, 1915. Indeed, on November
11th, Einstein submitted a note to the Prussian Academy of Sciences where
he wrote, essentially,[1] the equations $\mathbf{D}(\mathbf{g}) = \varkappa\mathbf{T}$ which carry his name. At
that stage, he did not have at his disposition any experimental confirmation
of his theory. However, in the following days (and before the 18th, the date

at which he submitted his results to the Academy), not only did he show that his theory predicted a deflection of light by the Sun two times greater than what he had himself previously predicted, but he showed that it explained an anomaly which had been observed for a long time and which had never received a satisfactory explanation.

On September 12, 1859, the French astronomer Urbain Jean Joseph Le Verrier submitted to the Parisian Academy of Sciences the text of a letter written to Hervé Faye in which he summarized some of his results. Le Verrier was already famous for the theoretical prediction he had made, in August 1846, of the existence of a new planet, Neptune. This planet was observed soon after, on the evening of September 23rd, in exactly the position in the sky which Le Verrier had calculated. In the years which followed, Le Verrier undertook an ambitious program, that of developing, for the first time, a global theory of the motion of the entire solar system. It is in developing this theory, and in comparing it with all available observations, to fix in a precise way all the free parameters in the problem (notably the values, unknown *a priori*, of the planetary masses), that he discovered a "grave difficulty," capable of putting into doubt Newton's law of gravity. This difficulty concerned the speed of rotation, in space, of the major axis of the ellipse traced by Mercury in its motion around the Sun.

We recall that, according to Kepler, the planets orbit around the Sun on elliptical trajectories. Newton's law of gravitation recovers this behavior in the approximation where one considers only one planet at a time, completely neglecting the presence of the other planets. However, as soon as one accounts for the perturbing forces exerted by the other planets on any one of them, one finds that its motion is perturbed and, in particular, that the ellipse on which it orbits the Sun is not fixed within space, but slowly rotates around the Sun. Astronomical observations had shown that it does indeed behave in this way. Since the perturbing forces acting on a planet depend on the value of the masses of the other planets, one may hope (if Newton's law is correct) to determine in a coherent fashion the values of the masses of all the planets such that they explain all of the observed perturbations in the elliptical motion of the planets. Such was the vast program to which Le Verrier consecrated more than a dozen years of work. Le Verrier found that he could determine, in a precise manner, the values of the planetary masses such that all of the orbital perturbations were explained, with one exception: the major axis of the orbit of the planet closest to the Sun, Mercury, advanced a little too rapidly around its center. Le Verrier could explain around 93% of the observed advance, but there remained an unexplained residue: a supplementary advance equal to about 38 arc-seconds per century. Note that this is quite a small shift in the advance. In one century, the unexplained residue is only on the order of magnitude of the angle spanned by one hair at the distance of one meter. However, for Le Verrier, it was clear that this effect was real. Although small in absolute value, it was quite significant in relative value since it corresponded to a modification of the mass of Venus by more than 10%, which was excluded by the other observational data. For some time, Le Verrier hoped to explain this effect by assuming the existence of another planet even closer to the Sun than Mercury. But this explanation, along with other suggestions, was abandoned since none of them could be confirmed by observation, and most had unacceptable consequences in relation to other existing observations.

No convincing explanation of the excessive advance of Mercury's major axis was found for more than 50 years. On the contrary, the simultaneous refinement of both observation and planetary theory only confirmed Le Verrier's discovery, and made the value of the still-unexplained advance more precise: at the beginning of the twentieth century, it was estimated at around 43 arc-seconds per century.

Einstein knew that any theory of gravity which differed from that of Newton was going to produce a certain supplementary advance of the major axes of planetary orbits. He also knew that, in a relativistic theory such as the one he had proposed, this advance was only going to be notable for the planet closest to the Sun, Mercury. Indeed, it is close to the Sun that the deformation of space-time is the greatest, and thus it is there that the theory has the most significant effects. He thus threw himself (between the 11th and 18th of November) into the calculation of the motion of a planet according to his theory. This is a relatively difficult calculation.

First, one must understand how a deformed space-time determines the world-line of a planet. Einstein had already understood in 1912 that his *principle of equivalence* dictated that a planet wound through space-time in a fashion such that its world-line would be as "straight" as possible, which is to say (in space-time) as "long" as possible.[2] In 1913, in collaboration with his dear friend Michele Besso, he had completed part of the calculation of planetary motion.

However, the most difficult part remained to be done: calculating the metric tensor **g** produced by the Sun. For that it was necessary to solve the equations he had written on November 11th. These equations are, *a priori*, quite complicated. Einstein succeeded in calculating, to the second order of approximation, the way in which the mass of the Sun deformed the chrono-geometry of the space-time around it. By combining these results, he could finally obtain the value predicted by general relativity for the anomalous advance of Mercury's major axis. Amazingly, he found 43 arc-seconds per century, precisely the amount which had remained unexplained! As he told his friends, this discovery gave him heart palpitations, and rendered him happily ecstatic for several days.

Einstein had often compared Nature to a Sphinx, who poses riddles to mankind, but almost never gives responses. In this case, Nature had just spoken to him, had told him: "Yes: the idea that mass-energy deforms the geometric structure of space-time is able to explain automatically that which has remained unexplained for such a long time." Einstein was convinced that general relativity had lifted a corner of the great veil.[3] He did not doubt that the other predictions made by general relativity would eventually be confirmed. However, as we have seen in the preceding chapter, the majority of physicists were not convinced until 1919, when the observations of the solar eclipse directly verified a *second* specific prediction of Einstein's theory: the fact that light rays were also deflected by moving through a region of space-time deformed by the Sun, while following world-lines which were as straight as possible.

Vibrational Waves in the Space-Time Gelatin

Another quite enlightening example of the new possibilities for "play" offered by the Einsteinian world is what one usually calls gravitational waves. Conforming to the image of space-time as an elastic gelatin, one may compare these waves to the waves which travel in the interior of a block of gelatin when one vibrates it. Let us note that one can vibrate a block of gelatin in many ways: either by shaking a material fiber embedded within the block, or by placing alternating stresses on the external surfaces of the gelatin. Einstein understood, as early as 1916, that these two processes were also possible in the case of space-time gelatin: either the mass-energy distribution within space-time shakes and thus provokes a vibrational process in the chronogeometry (this is the case, for example, when two stars rotate around each other, tracing a double helix in space-time); or a vibrational wave where the oscillating geometric structure of space-time arrives from infinity, propagating thanks to the elasticity of the space-time gelatin, and departs again to infinity.

Einstein was the first to submit these two possibilities to mathematical calculations. He showed, in 1916 and then again in 1918, that the theory of general relativity indeed admitted the existence of waves of space-time vibration. He found that the speed of propagation of these waves was exactly the same as that of light, 300,000 kilometers per second. This is much greater than the speed of propagation of elastic waves in ordinary solid media. For example, the speed of waves of elastic deformation in steel is 5 kilometers per second. Intuitively, the very great speed of waves of space-time deformation is due to the extreme rigidity $(1/\varkappa)$ of space-time, or in other words that the coefficient of elasticity $\varkappa$, discussed above, is so small.

Einstein also calculated the amplitude of gravitational waves emitted by a moving distribution of stress-(mass-)energy. He understood that these space-time waves themselves carried energy and momentum. From this he deduced that a moving distribution of stress-energy would lose energy because it radiates gravitational waves to infinity, and he obtained the expression giving this loss of energy, to the lowest order of approximation.

For a long time, it was believed that the process predicted and calculated by Einstein[4] produced a loss of energy too weak to be observable in reality. Indeed, if one estimates the radiation of energy, by gravitational waves, of sources that can be constructed on Earth (for example, a cylinder of several tons turning at the highest speed possible before breaking) one obtains a loss of energy that is ridiculously small. This situation only changed in the 1970s with the discovery of new astrophysical objects capable of condensing an enormous mass into a relatively quite small volume.

Of particular importance is the discovery by the American astronomers Russell Hulse and Joseph Taylor, in 1974, of the binary pulsar PSR 1913+16. This is a binary system made of two neutron stars which orbit, around the center of mass of the system, on highly elliptical trajectories. In this system, the loss of energy by gravitational radiation is large enough to have an observable effect. In fact, the best way to describe what has been observed for this system is the following. In November 1915 Einstein had verified that, to the lowest order of approximation, the theory of general relativity predicted that the gravitational interaction between two massive objects (through the deformation of the space-time between them) is described by the usual Newtonian law $F_{\text{Newton}} = Gmm'/r^2$. However, general relativity predicts modifications of Newton's law at higher orders of approximation. Roughly, these modifications depend on the ratio v/c between the orbital speed and the speed of light. The calculation of these modifications is complex. The first correction to Newton's law, proportional to the ratio squared v^2/c^2, was obtained[5] in 1917. After the discovery of binary pulsars, it was realized that it was necessary to push the precision of the calculation to a much higher level: up until the fifth power of the ratio v/c.

The complete result, of the form $F_{\text{Einstein}} = F_{\text{Newton}}(1 + v^2/c^2 + v^4/c^4 + v^5/c^5)$, for the Einsteinian interaction between two neutron stars, was obtained[6] in 1982. Among the new effects included in this Einsteinian interaction, the terms of order v^5/c^5 play a particular role. Their calculation shows that they come from *that part of the gravitational interaction which propagates between the two objects at the speed of light*. In other words, these terms reflect the existence of gravitational waves. By studying the effect of these terms on the orbital motion of a binary pulsar, one finds that they cause a progressive acceleration of the orbital frequency of the system, that is to say a progressive diminishing of the orbital period. For the binary pulsar PSR 1913 + 16, whose orbital period is around eight hours, this diminishment of the orbital period is, according to the predictions of Einstein's theory, 67 billionths of a second per orbit. Thanks to some very careful observations, collected over years, it has been possible to measure the diminishment of the orbital period of PSR 1913 + 16, and the observational result is in perfect agreement, within a few tenths of a percent, with the theoretical prediction. This agreement is one of the most beautiful confirmations of Einstein's theory. It is also the first confirmation that deformations of the space-time gelatin propagate (here, between the two neutron stars) at the speed of light.

In the 1960s, some scientists, including notably the American physicist Joseph Weber, understood that it was possible in principle to detect, on Earth, the arrival of gravitational waves which were emitted from very far away in the universe. A gravitational wave is a wave of deformation of the space-time geometry, which propagates from the source at the speed of light. Because of the very great rigidity of the space-time gelatin, all conceivable sources (including the most violent possible, like the fusion of two black holes) create extremely small deformations of the space-time geometry. However, to better understand in practice what a wave of deformation of the geometry might look like, we can follow George Gamow[7] and imagine the case of a gravitational wave with an amplitude so enormous that it would be detectable by a human being. On Earth, we are used to describing the geometry of surrounding space as an ordinary Euclidean geometry, where the Pythagorean theorem is true for right triangles of whatever size and where the sum of a triangle's angles is always equal to two right angles. Starting from this nondeformed or "flat" situation, let us see how Gamow describes the sudden arrival of a gigantic wave of geometric deformation at a British sea-side resort. "The professor," a scientist with a white beard, and Mr. Tompkins are installed in the lobby of the hotel, for an expository discussion of general relativity, while Maud, the professor's daughter, is practicing her painterly talents a little farther along the beach. Suddenly:

> While the professor was talking, very unusual changes seemed to be taking place around them: one end of the lobby became extremely small, squeezing all the furniture in it, whereas the other end was growing so large that, as it seemed to Mr. Tompkins, the whole universe could find room in it. A terrible thought pierced his mind: what if a piece of space on the beach, where Miss Maud was painting, were torn away from the rest of the universe. He would never be able to see her again!

The theoretical calculations, in general relativity, for the emission of gravitational waves by known (or conjectured) astrophysical sources show that what worried Mr. Tompkins has no risk of occurring. In fact, we are at every moment and in every location criss-crossed by waves of deformation of the geometry (or, for short, geometric deformations). But the amplitude of these waves is unimaginably small. The greatest geometric deformations that one could hope to see arrive on Earth (one or two times per year) would have an amplitude on the order of 0.000,000,000,000,000,000,001, or ten to the power minus 21. This means that the arrival of such a wave in the lobby of the hotel hosting Mr. Tompkins and the professor would shrink the width of the lobby by 0.000,000,000,000,000,000,000,1%, and would grow in length by the same proportion. It is clear that such a small effect is not visible to the naked eye!

The experimental physicist Joseph Weber was the first, near the end of the 1950s, to have the vision that modern technology might detect such small waves of geometric deformations. Today, after nearly a half-century of technological development, we hope to detect such waves in the coming years. In particular, the United States (the LIGO project) and Europe (the VIRGO and GEO projects) have recently completed the construction of gigantic interferometers, with arms that are kilometers in length, which have the potential of detecting such deformations. This immense technological effort is accompanied by an intense international theoretical effort to calculate the form of gravitational waves emitted by various astrophysical sources.

For example, one of the most studied and most promising sources is the "coalescence" of a system of two black holes orbiting around each other. We

have seen above that the light-speed propagation of the gravitational interaction between the two bodies of a binary system produces a progressive acceleration of the orbital frequency, itself connected to a progressive approach
of the two bodies towards each other. This effect has been experimentally
verified for several binary pulsars. After accumulating for hundreds of millions of years, this progressive approach of the two bodies leads to them being
so close together that they orbit around each other at speeds approaching
that of light. At this point, the progressive approach of the two bodies becomes more and more marked, and the orbits of the two objects become two
intertwined, shrinking spirals until the Einsteinian gravitational interaction
becomes so strong that it makes the two objects "fall" into each other. In the
particularly important case of two black holes, this spiraling descent of each
black hole towards the other leads to a fusion of the two black holes into a
more massive, rapidly rotating black hole. If Mr. Tompkins were situated in
the immediate vicinity of such a system of two fusing black holes, he could
witness a deformation of the geometry on the order of 10%, which is large
enough to be visible to the naked eye.[8] However, such systems being quite
rare in the universe, from Earth we expect only to see signals emitted by systems situated in galaxies very far away, on the order of hundreds of millions
of light years. Because of this, since the amplitude of the gravitational wave
decreases, as it propagates, as the inverse of the distance to the source, one
can only hope to detect minuscule deformations on Earth, on the order of
magnitude mentioned above.

Thinking of the Whole

Another example of a world-game rendered conceivable and mathematically
tractable by Einstein's theory is that of cosmology. The term *cosmology* existed
before Einstein, but Einstein has conferred to this word a new meaning, infinitely more ambitious than anything that had been thought of before. To
understand why a cosmological (that is, global) vision of reality was a central aspect of his vision of general relativity, we quote the way in which he
summarized what he saw as the essence of this theory, in a letter written on
January 9, 1916 to Karl Schwarzschild:

> The essential characteristic of my theory is precisely that no property is at-
> tributed to space in itself. One may express this in jest in the following
> manner: if every thing had just disappeared from the world, then, for New-

ton, there would remain the inertial, Galilean space; however, according to
my understanding, *nothing* would remain.

Today, one would be led to add some nuance to this assertion, since it has
been understood that the equations of general relativity also admit solutions
in the absence of all matter. Not only is Minkowski space-time one such
solution,[9] but there also exists a continuously infinite number of solutions,
describing vibrational waves in a space-time emptied of matter, which come
from infinity and leave towards infinity without having been created by any
matter whatsoever. However, the grandeur of Einstein's new understanding
can be seen in what it led him to conceive. Indeed, Einstein is the first to
think of matter-force and space-time as an indivisible whole. This indivisible
whole is called the *cosmos* (in the modern, Einsteinian sense).

In February 1917, Einstein wrote the article which laid the foundations
for all of twentieth-century cosmology, and which proposed the first mathematical model of the cosmos. It is difficult to overestimate the importance of
the conceptual leap that this article represents. Some modern authors treat
this article with condescension by remarking that the article missed the opportunity to predict the expansion of the universe. Indeed, amongst several
simplifying hypotheses, Einstein supposes that the cosmos is static. Since
he found that this hypothesis was incompatible with his other hypotheses
(a spatially homogeneous, closed universe with constant positive curvature;
the presence of a matter with a uniformly spread, positive mass-energy, but
without stresses), he modified his recently derived equations of general relativity by the addition of a supplementary term, called the *cosmological constant.* The addition of this cosmological constant permitted him to propose
the first global model of reality: Einstein's static cosmos. Soon after, other
authors, notably the Dutchman Willem de Sitter, and the Russian Alexander
Friedmann, understood that other models for the cosmos were possible, and
that the most general cosmos should be not only "curved in space," but also
"curved in time," that is to say either expanding or contracting.[10] It was then
understood that the modification of general relativity by a cosmological constant was not necessary, when one considered a cosmos filled with matter and
curved in time.

We know the remarkable scientific offspring of such a cosmos. Through
the observational work of the American astronomers Vesto Slipher (in 1915)
and Edwin Hubble, and the theoretical work of Georges Lemaître and George
Gamow, it led to the model of the Big Bang which was confirmed by the

discovery of the background cosmological radiation and the explanation of the cosmological density of light elements (deuterium, helium, and lithium). This model was then completed by the idea of a primordial *inflationary phase* and by the recent observational discovery that the cosmos has just entered into a new inflationary phase. We here defer to the numerous works which exhibit in detail modern cosmology and its origins.[11]

In sum, we find that at the conceptual level the cosmological science of the twentieth century finds its foundation in the article written by Einstein in February 1917. To unite space-time and matter-force, *the container* and *the contained*, into one whole was an act of extreme intellectual courage. At any rate, Einstein was conscious of the audacity of his gesture. On February 4, 1917, he wrote to his friend Paul Ehrenfest that he had "again committed, in regards to gravity, something which puts him in danger of being shut up in an insane asylum." Today, proposals for the relativistic cosmos, which take into account a great number of observed facts, differ in most of their details from the original cosmos which sprang from Einstein's brain in 1917. Curiously enough, one of the details of Einstein's cosmos, the cosmological constant, which had long been considered as a "blunder" by Einstein, has been recently revealed to be an essential constituent of the present universe. Under the new name of *dark energy*, it seems today that it represents around 70% of the distribution of stress-energy in the universe.

To conclude, we comment briefly on the notion of "cosmic time" in the relativistic cosmos. Popular treatments of science have a tendency, when speaking of cosmology, and especially of the Big Bang, to use a language which suggests that general relativity reintroduces that notion of temporal flow which special relativity had abolished. No such thing. The space-time of general relativity is just as "immobile" as that of Minkowski. The Big Bang is not the "birth" of the universe, or its creation *ex nihilo*, but one of the possible boundaries for a strongly deformed space-time. In the analogy between Einstein's equations and the equations of elasticity, one could say that a big bang (or a big crunch, which is the same thing, seen "up-side down"[12]) is the result of having passed the threshold of elasticity of the space-time gelatin and of having attained the regime of rupture. A big bang is thus like the end of a broken rubber band.

Far from reintroducing the notion of temporal flow, the infinite variety of possible Einsteinian cosmological models furnish some striking examples of "worlds" where the unreality of this flow becomes palpable. For example, among these possible[13] cosmologies, one can imagine a space-time where

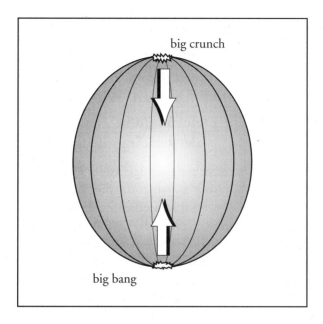

Figure 11. A possible cosmos where time does not everywhere "flow" in the same direction.

there are both big bangs and big crunches, such that the privileged arrow of time[14] (for thermodynamics[15]) in the vicinity of these various space-time boundaries is, for each boundary, directed towards the interior of the space-time (as it is for the boundary of our space-time, conventionally called the Big Bang). In such a cosmos, conscious beings living in different domains of the same space-time (let's say close to a "lower" boundary or close to an "upper" boundary) would find that time "flowed" in mutually contradictory directions: that which is future for one is past for the other! Another example of a relativistic cosmos which puts into question the usual notion of temporal flow is the one introduced in 1949 by the famous mathematician (and Einstein's colleague at the Institute for Advanced Study) Kurt Gödel. In Gödel's cosmos, time can turn around on itself. Indeed, there exist world-lines, representing the history of observers living in this cosmos, which close in on themselves like circles. An observer living along one of these lines would experience Nietzsche's eternal return, which is to say he would live his life as a loop (meaning that he would live a finite life where the future loops into the past), while an observer living along an infinitely long world-line would experience a linear time without return.[16]

In sum, these examples of relativistic cosmologies have features that would have given Bergson nightmares. However, they also provide very rich food for thought, concerning the notion of time and the philosophical impact of Einstein's discoveries.

Large Space-Time Deformations: Neutron Stars and Black Holes

To finish our survey of the new horizons opened up by the theory of general relativity, we describe what happens when the distribution of stress and energy is so concentrated that it leads to a quite significant deformation of the chronogeometry of space-time. This is the case for neutron stars and black holes, which are two of the possible final states for massive stars. Recall that most of a star's life is spent slowly burning its nuclear fuel. This process causes the structure of the star to take the form of a series of layers of differing nuclear composition, surrounding a core that becomes more and more dense. When the initial mass of the star is sufficiently large, this process finishes by leading to a catastrophic event: the core, already much denser than ordinary matter, collapses in on itself under the influence of its gravitational self-attraction. Depending on the mass contained in the core of the star, this collapse can lead to the formation of either a neutron star or a black hole.

A *neutron star* condenses a mass approximately equal to the mass of the Sun into a radius of about 10 kilometers. The matter comprising such a star has a nuclear composition dominated by neutrons (the protons and electrons having reacted together to transform into neutrons, with an associated emission of neutrinos). The density of mass-energy in the interior of a neutron star exceeds 100 million metric tons per cubic centimeter. Moreover, the stresses in such a star (in the form of pressure of the neutron gas) become themselves so enormous that they contribute significantly to the space-time deformation. In solving Einstein's equations, it is found that a neutron star deforms the chronogeometry of space time much more than the Sun does.

Let us give some idea of the respective deformations caused by the Sun and by a neutron star. We recall that, if geometry were Euclidean, the sum of the angles in a triangle would equal 180°. An ordinary triangle is simply the spatial figure obtained by joining three points in space by straight lines. According to Einstein, the (spatial[17]) geometry in the vicinity of a distribution of stress-energy is no longer Euclidean. One may nevertheless continue to naturally define a triangle by joining three points in space by the shortest possible lines. Consider a triangle (contained in a plane passing through the center of the object) which just barely encloses the star considered (the Sun or a neutron star), in other words whose sides just barely touch the surface of the star. One measure of the deformation of the geometry will be to compare the sum of the angles of such a triangle circumscribing the star to the value (180°) that it takes in the nondeformed Euclidean case. In the case of the Sun, the sum of the angles of such a triangle is larger than 180° by about three arc-seconds. The relative deformation (three arc-seconds divided by 180°) is only four millionths. Quite a small deformation of the geometry! On the other hand, the sum of the angles for a triangle circumscribing a neutron star is larger than 180° by about 70°. The relative deformation is now around 40%! We now see in what sense a neutron star creates a large deformation of the geometry. One may also conclude that if one could make observations which confirmed that general relativity correctly describes the gravitational field created by a neutron star, one will have also verified that this theory is useful in the regime of large space-time deformations. Without entering into details, we shall mention only that four different binary pulsars have allowed for *ten independent confirmations* of general relativity in the regime of large space-time deformation. Four of these confirmations verify at the same time the reality of the propagation of gravitational waves, such as is predicted by general relativity. We note finally that some of these confirmations attain a remarkable precision, with a relative error on the order of three thousandths. We may add that a very large number of observations of the solar system (which complete the historic test of the advance of Mercury's major axis) have verified the predictions of general relativity, in the regime of weak deformations of the chronogeometry, with a precision which is at the least on the order of three thousandths, and which in one case attains the exceptional level of two hundred-thousandths (2×10^{-5}).

All of these highly precise observational verifications (as well as many others that we shall not mention here) make general relativity one of the best confirmed theories in modern science. This is why it is justified to take seriously the entirety of this theory's predictions, even if they have still not received direct confirmations. This is the case for predictions concerning the limit of space-time deformations which are even greater than in the case of neutron stars, so much so that they pass the threshold of elasticity of the space-time gelatin. In the case of ordinary elastic media (gelatin, a block of rubber, a piece of metal), when one applies stresses that are too strong, one passes out of the regime of elasticity (which is reversible, i.e., such that the body returns to the nondeformed state when one stops applying the stress) and successively enters (i) into the *regime of plasticity* (where the body deforms in an *irreversible* fashion, but without breaking), then (ii) into the *regime of rupture* (where the body breaks or tears). These two regimes have analogs in the case of space-time elasticity. One could say that the formation of a *black hole* corresponds to the regime of plasticity of the space-time gelatin. Then, one may compare (as we have already indicated) the formation of *cosmological singularities*[18] (big bang or big crunch) to a tearing of the space-time gelatin.

A black hole is the result of the continued collapse of a star, that is to say a collapse which does not stop with the formation of a neutron star. The concept of a black hole emerged very slowly from general relativity. In January 1916, the German physicist Karl Schwarzschild succeeded in finding the first exact solution of Einstein's new equations. Schwarzschild believed this solution to represent the exact form of the space-time deformation created by the Sun (the same deformation calculated by Einstein in November 1915, but only to the second order of approximation). However, the exact solution that he found exhibited some strange and surprising behavior close to its center. This strange behavior is connected to what is today called the black hole horizon or the surface of a black hole. It took nearly 50 years of subsequent work on this strange behavior in order to understand its conceptual impact. We shall not retrace here the slow birth of the concept of the black hole.[19] We shall content ourselves here to highlighting three important stages. The physical concept of the black hole, as the result of the continued collapse of a star, was introduced by John Robert Oppenheimer and Hartland Snyder in 1939. The global chronogeometric structure of black holes was only understood in the 1960s through a series of works, notable among them those of Martin Kruskal and of Roger Penrose. The name *black hole* was introduced by John Archibald Wheeler in a lecture given on December 29, 1967.

The reader will find in Figure 12 the schematic representation of the space-time chronogeometry of a black hole formed through the collapse of a spherical star.[20] This diagram represents a three-dimensional space-time, with two spatial dimensions and one temporal dimension. The circle, or rather the disk, on the bottom of the diagram represents the initial state, considered at a certain "zero" instant, of a star in a two-dimensional space. In the future, towards the top of the diagram, this star collapses and thus takes the form of a succession of disks whose radii become smaller and smaller. The spatio-temporal stack formed from these successive disks traces the space-time history of the collapsing star. This collapse creates a distribution of mass-

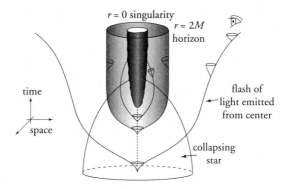

Figure 12. Black hole formed by the collapse of a star.

energy-stress which becomes more and more dense, thus causing more and more deformation of the space-time chronogeometry. To simplify the visualization of the latter, we have not indicated the hourglasses (representing the events separated from each point by a small, positive squared interval), but the *light-cones* (representing the events separated from each point by a squared interval equal to zero). Moreover, we have only kept the upper part of the light cones, opening towards the future. Each cone represents the (infinitesimal) beginning of the space-time history of a flash of light emitted, in all directions, from a certain point in space at a given instant. More generally, one may associate with each event its future (deformed) light conoid, which is the figure defined by the complete (and thus no longer infinitesimal) space-time history of a flash of light emitted from this event. In sum, such a conoid is the history of a *bubble of light* which expands outward from an initially vanishing radius. The interior of this future conoid is the future of this event, that is to say the part of space-time that it can influence or to which it can send information. Many of these conoids (in the form of tulips or wine glasses opening towards the top) are represented in the diagram of Figure 12.

The essential feature shown by this diagram is the formation of a region of space-time (the larger, shaded zone) from whence nothing can escape: neither light, nor matter, nor information. The frontier between this shaded zone (called the "black hole interior") and the clear external zone is called the black hole horizon or the surface of the black hole. Those cones whose apexes are situated in the clear zone (the black hole exterior) will evolve into conoids which escape (at least partially) out to infinity, representing the possibility of emitting, from this clear zone, electromagnetic signals towards infinity. On the other hand, the cones whose apexes are located in the shaded zone (the black hole interior) will evolve without ever being able to leave the shaded zone. It is therefore impossible to emit, from the shaded zone, an electromagnetic signal that reaches infinity. From this comes the adjective "black" used to describe this structure. We note, however, that the surface of a black hole, or the horizon, far from being black, is, in fact, constructed from the history of a *light bubble* which departed, at a particular moment, from the center of the star, which started to escape, and which is then stabilized in the form of a space-time cylinder. This space-time cylinder (the upper, stationary part of the horizon) thus represents the space-time history of a light bubble which, seen locally, moves outward at the speed of light, but which globally runs in place. This remarkable behavior illustrates the fact that, in a black hole, the stresses exerted by the distribution of matter have exceeded the elastic limit and have reached the regime of plasticity, where the space-time gelatin begins to resemble a fluid which flows into a hole. Indeed, one could compare the bubble of light which jogs in place to what may happen around an opened drain in an emptying sink: a wave could move through the water, away from the drain, all the while remaining stationary with respect to the sink, through the effect of the motion of the water towards the drain.

We note another very important element in the structure of a black hole. The temporal development of the interior region is limited, terminating in a space-time boundary (the dark gray surface) where the deformation of the chronogeometry (in the sense of the curvature tensor) becomes infinitely large. Space-time ceases to exist beyond this boundary, which is found to be a big crunch, the temporal inverse of a big bang. In our analogy with elastic media, this boundary is similar to the place where an elastic medium breaks. In other words, the interior of a black hole contains an anticipated "end of time" where the fabric of space-time tears.

Let us point out that besides being a global chronogeometric structure, it is also fruitful to consider a black hole as being a physical object localized in the external space and persisting in time: in other words as a sort of dead star, tracing a tube in space-time. The space-time tube traced by a black hole is nothing but its horizon, also called the surface of the black hole, represented by the outer gray cylinder in the figure. In studying the physics of these objects, it has been shown that they can be attributed with many of the properties of ordinary bodies, for example: a mass, an energy, a momentum, and an angular momentum.[21] Moreover, it can be shown that black holes can exchange energy, angular momentum, and electric charge with their environment. Demetrios Christodoulou and Remo Ruffini have even shown[22] that black holes are the best reservoirs of free energy in the universe: indeed, 29% of their mass energy (mc^2) can be stored in the form of kinetic energy of rotation, and up to 50% in the form of electric energy. These percentages are much higher than the energies (representing a few percent of the mass energy) which can be extracted from the nuclear fusion reactions which are the origin of all the light emitted by stars during their lifetimes. In addition to their mechanical properties (energy, momentum, etc.), it is also fruitful to attribute thermodynamic properties (entropy[23] and temperature[24]) to black holes, and even dissipative properties localized on their surfaces (surface electric resistivity[25] and surface viscosity[26]).

Although there is, at this moment, no irrefutable proof of the existence of black holes in the universe (despite some media-friendly announcements, which only concern processes occurring quite far from the horizon of an eventual black hole), there is an abundance of indirect evidence which points in favor of their existence. In particular, more than a dozen x-ray–emitting binary systems in our galaxy are probably made up of a black hole and an ordinary star. Moreover, the center of our galaxy seems to contain a very compact concentration of mass, equivalent to that of three million suns, which is probably a black hole. The detection of gravitational waves emitted from the coalescence of two black holes, which may hopefully occur soon, should supply direct, irrefutable proof of the existence of black holes, by exhibiting the characteristic frequencies of vibration of the final black hole formed from the fusion of two initial black holes. Indeed, it can be shown that a black hole is an elastic structure which can vibrate, and that its vibrations cause space-time to vibrate around it, in a way completely analogous to a bell which vibrates, and whose vibrations excite sound waves in the surrounding air.

Einstein (standing, second from right) next to Langevin and behind Poincaré, at the first Solvay Council (Brussels, 1911). (*Credit Rue des Archives / TAL.*)

5

Light and Energy in Grains

Einstein would be one of the greatest theoretical physicists of all times even
if he had not written a single line on relativity.
—Max Born

It seems quite certain that from now on one must introduce into our
physical and chemical ideas a discontinuity, an element varying by jumps,
of which we had no conception a few years ago.
—Marcel Brillouin

"An Element Varying by Jumps"

Brussels, Belgium, October 30–November 3, 1911

On October 30, 1911, the Belgian industrial tycoon Ernest Solvay opened
the first Solvay scientific council in a small room of the Métropole, the most
beautiful hotel in Brussels. It was "a sort of private conference" to which
Ernest Solvay had convened the most prestigious scientists of the moment to
discuss, in a small group, the revolution which had been shaking the foundations of science. Participation in this mini-conference was limited to around
20 scientists, who had all received a personal invitation from Solvay. Around
half of them would later be awarded (or already had been awarded) a Nobel
prize.

Solvay was an autodidact who was absolutely enthralled by science. His
fortune came from an industrial application of chemistry (a procedure for the
cheap production of sodium carbonate). He was also a visionary of science,
who wanted and knew how to dedicate a part of his fortune to the promotion
of existing science and was thus able to assist the discovery of new scientific
ideas.[1] He also loved to meet scientists; his life's great regret was having had

to renounce the pursuit of his studies and his entry into the university because of sickness in his youth. As he wrote: "To be in contact with the scientist, to become a bit of a scientist myself if possible, to envisage the revision of physical results, to thus unveil the real, the definitive, was the golden dream of my entire life." His activity as a patron of science was a great success, first through the organization of the prestigious series of Solvay councils, which did so much to aid the development of new scientific ideas through intensive discussions, then by the foundation of the International Solvay Institutes of Physics, Chemistry, Physiology, and Sociology.[2]

In 1910, Ernest Solvay contacted the great German physical chemist Walther Nernst to ask for his help in organizing this scientific meeting on "some questions of present interest" in physics and chemistry. Nernst suggested that Solvay dedicate this first Solvay conference to the revolutionary changes which were demolishing the foundations of physics, previously considered unshakable. Nernst explained to Solvay that this revolution (called the quantum revolution) had developed from the work of his physicist colleague in Berlin, Max Planck, and (above all) from the work of Einstein. Nernst had in mind, in particular, a work on the specific heats of solids (see below), published in 1907. This work of genius was the first application of the idea of quanta outside of the domain of radiation, where their presence had initially been discovered by Planck and Einstein. Einstein's generalization of the quantum ideas had allowed him to account for the mysterious thermal behavior of certain solids (diamond in particular) which had defied all explanation for around 50 years.

Thus, during the week from October 30 to November 3, 1911, the first Solvay council took place in Brussels, entitled "The Theory of Radiation and Quanta." The list of the 20-odd participants is impressive. We here note only Hendrik Lorentz (Nobel prize in physics, 1902), Marie Curie (Nobel prize in physics, 1903, shared with Pierre Curie and Henri Becquerel; and Nobel prize in chemistry, 1911), Max Planck (Nobel prize in physics, 1918), Jean Perrin (Nobel prize in physics, 1926), Ernest Rutherford (Nobel prize in chemistry, 1908), Walther Nernst (Nobel prize in chemistry, 1920), Wilhelm Wien (Nobel prize in physics, 1911), as well as Henri Poincaré, Paul Langevin, Marcel Brillouin, Maurice de Broglie (elder brother of Louis de Broglie), and of course, last but not least, Albert Einstein (Nobel prize in physics, 1921). Einstein was the youngest of the participants (at 32 years old), as one may see in the famous photograph from the first Solvay council. Einstein, visibly quite relaxed, is nonchalantly smoking a cigarillo next to Langevin. They are both standing close to a seated, aging Poincaré, who seems to be explaining a mathematical result to an attentive Marie Curie.

Let us return to the morning of October 30, 1911, when the first Solvay council opened at the Métropole hotel. After a few words of welcome, Ernest Solvay opened the meeting with a brief speech in which he made some remarks about his own theory of space, matter, and energy. Nevertheless, he admitted that this theory belonged more to physical philosophy than to current physics. He then passed the floor to the chairman of the meeting, Hendrik Lorentz. It would have been difficult to find a better chairman than Lorentz. Born in 1853, and having received, in 1902, one of the very first Nobel prizes in physics (the first was awarded in 1901 to Wilhelm Röntgen), Lorentz was one of the "wise men" of physics. In addition to his native Dutch, he could speak fluently in French, German, and English, which allowed him to preside with ease over the progress of the meeting, which took place in these three languages (a happy time of linguistic diversity!). Lorentz acquitted himself of his task as chairman with tact and diplomacy. For his entire life, Einstein had a very deep respect for him. He admired Lorentz's harmonious personality, and had an almost filial affection for him. It is perhaps surprising that the exceptional human resonance between Einstein and Lorentz was able to overcome their scientific ideas, which were often directly opposed to each other. For example, the conservative Lorentz held to the concept of ether and never believed in relativity in Einstein's sense, and was moreover attached to the description of light by a continuous electromagnetic field, while in March 1905 Einstein had proposed the "revolutionary" idea of a discontinuous structure for light (see below). This may be sensed in Lorentz's inaugural speech, delivered on the morning of October 30, 1911, in which he names the

> important questions which we shall have to occupy ourselves with. I say *important*, because they bear on the very principles of mechanics, and on the most intimate properties of matter. Perhaps, although let us hope that it will not be thus, even the fundamental equations of electrodynamics, and our ideas on the nature of the ether, if it is still permitted to use this word, shall find themselves a little compromised.

He continued in a more positive register by saying:

> [...] We now have the feeling of finding ourselves at an impasse, the old theories being shown to be more and more useless for piercing the shadows which surround us on all sides.

> In this state of things, the beautiful hypothesis about the elements of energy, written for the first time by Mr. Planck and applied to a number of problems

by Mr. Einstein, Mr. Nernst, and others, has been a precious ray of light. It has opened up to us unexpected perspectives, and even those who look on it with a certain mistrust must recognize its importance and its fecundity. It thus merits well to be the principal object of our discussions, [even if] this new idea, as beautiful as it is, raises in its turn serious objections.

In sum, the focus of the discussions held at the Métropole hotel in this autumn of 1911 had come (apart from the inaugural work by Planck in 1900 which had not had any great impact in the community of physicists) from the innovative work completed in the period from 1905 to 1907 by the youngest of the participants in the meeting: Albert Einstein. Indeed, Einstein had been given the role of being the star attraction, and was scheduled to give the final lecture.

What did Einstein, who had finally received a full professorship (in Prague), think of this first Solvay council, the first international conference which he had attended? On one side, he appreciated the opportunity to have detailed discussions with his colleagues, and to try to convince them to re-nounce their old mental habits. For example, he wrote in his letters: "Planck is blocked by some indubitably erroneous prejudices . . . but no one sees clearly in all this. There should be in this whole business enough to please a com-pany of diabolical Jesuits," as well as: "I have succeeded in convincing Planck into admitting a good number of my ideas, which he has resisted for years. He is a man of fundamental honesty, who thinks of others before thinking of himself."

On the other side, in December 1911 he wrote to his most intimate friend, Michele Besso: "This meeting has all the appearance of a lamenta-tion on the ruins of Jerusalem. Nothing positive has come of it. My own hesitant contributions raised great interest with no serious criticism. I have profited very little, since everything I have heard was already known to me."

Even if Einstein learned nearly nothing, his colleagues understood that a new world had opened up in physics, that of quanta, and that Einstein was its most intrepid pioneer. The quote from Marcel Brillouin, placed at the start of this section, expresses well the sentiment of all the participants at the meeting that quantum discontinuity was there for good. And the general public? Would it play its part in the excitement caused by the revelation of this new world? Would the press mobilize itself to speak of this memorable scientific event? Yes, the first Solvay council hit the presses, but not for sci-entific reasons! The French press, followed by the Belgian, openly attacked Marie Curie at the end of the meeting for the "guilty liaison" that she participated in with Paul Langevin, a husband and father of four children. (Pierre Curie had died in 1906.) The two scientists were both present in Brussels, which rendered the scandal more manifest. This xenophobic and sexist press campaign (Madame Curie was pictured as a *métèque*, a derogatory way to refer to a foreigner, and as a *fatale étudiante*, a variation on the more familiar term *femme fatale*), was above all meant to keep Marie Curie away from the French Academy of Sciences (with success, at any rate). Without this scandal, the readers of the European press would doubtless not have heard talk of this meeting, which marked an essential step in the appreciation and diffusion of the quantum revolution. Here, we will simply take note of two of the repercussions of the animated discussions held in Brussels: the fact that Niels Bohr was completely engrossed when Rutherford, after returning to Manchester, gave him a "striking account of the discussions at the first Solvay council," and that Louis de Broglie was seized with a passion to bring to light the mystery of quanta after reading the edited accounts of the meeting (in French) by his older brother, Maurice de Broglie.

Now, let's go back through the years, and explain the path of thought by which Einstein succeeded in convincing part of the community of physicists of the existence of discontinuities in those domains of physics where the continuous had reigned as master for so many years.

A "Very Revolutionary" Idea

Bern, Switzerland, March 1905

We now return to the "miraculous year," 1905, when a young employee of the Bern Patent Office laid the foundations of modern physics. We always associate Einstein with the theory of relativity, or more exactly to the *two* theories of relativity, but as clearly expressed in the statement by the German physicist (and Nobel laureate) Max Born quoted at the beginning of this chapter, even if Einstein had written nothing on relativity, his other contributions to theoretical physics are so fundamental that they would suffice to make him one of the greatest physicists in the history of thought. So what are these other contributions? When did Einstein find time to make these fundamental contributions, he who was so busy with special relativity from 1905 to 1907, then with the construction and development of general relativity from 1907 to 1918? In fact, Einstein made his most revolutionary contribution to twentieth-century physics four months before his June 1905 article on special relativity. In addition, he continued to propose profoundly innovative ideas concerning the intimate nature of light and matter in a regular fashion between 1905 and 1924. These ideas, along with others due to Max Planck and Niels Bohr, are at the base of the third great scientific revolution of the twentieth century: the quantum theory, which found its first complete formulation in 1925 and 1926 in the works of Werner Heisenberg, Max Born, Pascual Jordan, Erwin Schrödinger, and Paul Adrien Maurice Dirac.

In March 1905, having just turned 26, Einstein put the final touches on an article proposing a new point of view on the nature of light, and of its absorption or production by matter. In a letter written soon afterwards to his friend Konrad Habicht, one of the three members (with Einstein and Maurice Solovine) of the Olympia Academy, he spoke of the contents of this article in these terms: "It addresses the question of radiation and of the energetic properties of light, and this in a very revolutionary fashion ... " Remember that Einstein never called his work from June 1905 on relativity revolutionary, but that he considered it only as a conceptual step. What, then, was the revolutionary idea of this article of March 1905? It was the suggestion that, contrary to what everyone considered to be firmly established at this epoch, luminous energy is not spread in a continuous fashion in space. Rather, it is concentrated in little *quanta of light* (*Lichtquanten*): grains of luminous energy localized at particular points in space.

The revolutionary character of this idea can be measured by the fact that it was rejected as absurd by nearly every physicist for 20 years! Even Max Planck, often considered to be the initiator of the quantum revolution, rejected Einstein's hypothesis of luminous quanta as an aberrant speculation. And yet, Planck had been the first theoretical physicist to recognize the grandeur of Einstein's innovative ideas, most notably his work of June 1905 on relativity. In 1913, Planck and some of his colleagues wrote a report on Einstein to propose his candidacy for the Prussian Academy of Sciences. In this report, they praise the exceptional importance of Einstein's contributions to physics. Nevertheless, they end up concluding their report on a negative note which tells a lot about their opinion of Einstein's "very revolutionary" idea:

> To summarize, one could say that among the great problems in which modern physics abounds, there are nearly none to which Einstein has not brought a remarkable contribution. It is true that he has sometimes missed the goal in his speculations, for example, with his hypothesis of luminous quanta; but he should not be reproached for it, since it is not possible to introduce really new ideas, even in the most exact sciences, without sometimes taking risks.

The Wave Nature of Light

Since the beginning of the nineteenth century, the nature of light had seemed well understood. The Englishman Thomas Young (between 1801 and 1807) and the Frenchman Augustin Fresnel (between 1815 and 1821) had proven, through innovative experiments, that light behaved like a wave. Ripples moving along the surface of a tranquil lake can, when they cross each other, produce what is called *interference*: an alternation between zones where the superposed ripples reinforce each other, and others where they cancel each other, and leave the surface of the water unmoved. Similarly, the experiments of Young and Fresnel had shown that light coming from a single source, but having traveled over different routes before being recombined, produced interference, now an alternation of bright and dark zones. From this they concluded that light was a wave-like phenomenon, whose energy was thus spread in a continuous fashion throughout space and which propagated, crest by crest, at a speed of 300,000 kilometers per second.

This conception of light was confirmed through the work of James Clerk Maxwell (around 1860) who proposed the identification of light as an *electromagnetic wave*, that is, as mentioned previously, an oscillatory phenomenon in which an electric field is transformed during its propagation into a magnetic field and vice versa. The experiments of Heinrich Hertz between 1886 and 1888 established the existence of electromagnetic waves moving at the speed of 300,000 kilometers per second, and verified their ability to be reflected, to be refracted, and to interfere, in a way perfectly similar to light waves. Because of this, after 1887 the answer seemed clear: light was an electromagnetic wave whose energy was spread continuously throughout space.

The Worm within the Fruit

As we have just said, it was the experimental work of Heinrich Hertz which established the existence of electromagnetic waves and seemed to confirm in a definitive way the wave nature of light. Nevertheless, in the course of his experiments Hertz accidentally discovered, in a strange irony of history, a new effect which was going to constitute, thanks to Einstein's work from March 1905, one of the most convincing proofs of the *corpuscular* nature of light. This new phenomenon is called the *photoelectric effect*, since it combines light (*photos* in Greek) and electricity. It is for the theoretical discovery of the fundamental law of the photoelectric effect that Einstein was awarded the 1921 Nobel prize in physics.

It is interesting to describe how this effect was discovered by Hertz in 1886, to show how a purely experimental discovery can appear incomprehensible, inasmuch as it is not included within the framework of an explanatory theory. Hertz produced electromagnetic waves by causing an electric charge to oscillate between two closely situated balls of copper, which were connected to an induction coil. An oscillating current in the induction coil produced an oscillation of the charge on the copper balls, and thus an electric arc was formed between the balls: a strong, glowing spark was created through the electric ionization of the layer of air between the copper conductors. This Hertz oscillator produced electromagnetic waves in the radio-wave range of frequencies (which were thus invisible to the human eye). To detect these radio waves, Hertz used a copper wire formed into a loop, with a small gap between its two extremities (of which one was sharpened to a point and the other rounded into a sphere). When the frequency of oscillation of the Hertz oscillator was tuned to the size of the receiving ring, Hertz detected the arrival of electromagnetic waves at the detecting ring through the formation of a tiny electric spark in the gap of the ring. Since this spark was difficult to see, Hertz had the idea of enclosing the receiving ring in a black box to be better able to see it. However, to his surprise, upon enclosing the receiver within a box, the spark became much weaker! After extensive experimental study, Hertz understood that the reason for this strange effect was to be found in the influence, on the receiving ring, of the ultraviolet light emitted by the electric arc which formed between the copper balls of the oscillator which was producing the radio waves. Hertz published his observations in 1887, but he could not supply any explanation of the mechanism by which illumination by ultraviolet light could influence the size of the spark appearing in the ring.

The photoelectric effect was studied experimentally for 30 years following Hertz's initial work. The discovery of the electron, around 1897, through the work of Jean Perrin and Joseph John Thomson, permitted workers to close in on the basic mechanism behind the photoelectric effect: the surface of a solid body (the copper, in the case of Hertz's experiments) illuminated by ultraviolet light emits electrons. But the most remarkable results on the photoelectric effect were obtained around 1900 by Philipp Lenard, a former assistant of Hertz. Indeed, it is Lenard who discovered some aspects of this phenomenon which seemed utterly incomprehensible from the perspective where light is considered as a wave, whose energy is continuously spread throughout space. For example, one of the most unexpected processes that Lenard discovered is the following: Lenard explored the emission of electrons by a solid under the

influence of ultraviolet light of various frequencies. He observed that when
the frequency of ultraviolet light became smaller than a certain threshold, the
illuminated surface stopped emitting electrons. Lenard raised the intensity
of the ultraviolet radiation, waited for a very long time, and still nothing!
No electrons escaped from the illuminated surface. And yet, the energy contained within the light wave arriving at the surface was more than sufficient
to extract electrons from the solid body and to supply them with kinetic energy. How could it be that the same amount of luminous energy, although
arbitrarily large, becomes totally incapable of extracting electrons from a surface, as soon as the frequency of oscillation of the light passed below a certain
threshold?

Einstein had read Lenard's work from 1901 with enthusiasm, when he
had just come out of Zurich Polytechnic. In the same epoch, he had also
read other works on the nature of light, and particularly the work of Max
Planck, published in 1900, which indicated that the exchange of energy between matter and light exhibited some bizarre features. The young Einstein
had an exceptional flair for locating those areas of physics where there were
new things to comprehend. In order to understand the path of thought which
led him to formulate his revolutionary hypothesis on the quanta of light, let
us go back and explain what Planck's foundational works were all about.

The Black and the Red

In everyday life, a "black" body is one that absorbs, without reflection, any
light which falls on its surface. In fact, since the spectrum of *visible* light, that
which is perceptible to the human eye, only covers a range of wavelengths between 0.4 and 0.8 microns (a micron being one thousandth of a millimeter),
a body which seems black to us could seem "infra-colored" or "ultra-colored"
to another being which is sensitive to infrared or ultraviolet radiation. In
physics, where we love the limiting cases which permit us to simplify our description of nature, we define a *black body* to be an ideal body whose surface
absorbs, without reflection or diffraction, all light which falls on its surface,
whatever its wavelength may be. The fact that a body is black, in the sense
of being a perfect absorber, does not forbid it as well from being an *emitter* of
light. For this to happen, it is enough to simply heat the body. For example,
common experience shows that a piece of metal which is black at ordinary
temperatures, becomes red when heated sufficiently, and even white if one

continues to raise its temperature. This variation in the color corresponds to the change, with temperature, of the emissivity of a "black body."

One way to construct a black body is to make a small hole in an otherwise closed oven. The surface of such a hole is black, in the sense of being a perfect absorber. Indeed, any light falling on this hole will penetrate into the oven, and will there experience multiple absorptions, reflections, and diffractions, such that it will eventually be completely absorbed by the oven's interior surface. In other words, it is a trap for all incident radiation. In consequence, one can study emission from a black body by studying the fashion in which the heat radiation inside the oven is distributed, for a given temperature, with respect to the frequency of the constituent electromagnetic waves.

The distribution of an oven's heat as a function of the frequency was much studied, both experimentally and theoretically, in Germany and Austria in the second half of the nineteenth century.[3] In 1896, Wilhelm Wien proposed a certain simple mathematical formula to represent the distribution of an oven's heat with respect to frequency, also known as the law of black-body radiation. For four years, it seemed that Wien's law gave a perfect representation of black-body radiation. However, in 1900, measurements made by two experimental groups in Berlin showed that Wien's law, which had been verified for relatively large frequencies of light, did not correctly describe the radiation emitted at small frequencies (or, equivalently, long wavelengths). These experimental results led the theoretical physicist Max Planck (who also worked in Berlin) to propose, on October 19, 1900, a new mathematical formula, slightly more complicated than that of Wien, to represent the measurements of his colleagues on the radiation of a black body. And indeed, his colleagues found that this law described their measurements almost perfectly.

Planck's law for black-body radiation contained the seed of one of the most profound revolutions in the history of science. Planck had obtained his law through an educated guess, based on a solid knowledge of black-body thermodynamics as it was understood at the time. He understood right away that his law must contain new information on the physics of interaction between matter and light. In the months following his conjecture, he used all of his mental resources in an attempt to derive his law from the fundamental laws of physics of his time, but he failed repeatedly. His greatness as a physicist is seen in the fact that his repeated failures did not discourage him from moving into unknown territory. For years he had focused his research on the law of black-body radiation, because he knew (as Wilhelm Wien had shown) that this law was one of the rare universal laws of physics. There are indeed very few universal laws in physics, and he thus wished to reach a deep understanding of what this one was saying about nature. As he later wrote:

> For six years, I beat myself up with the theory of black bodies. I had to find a theoretical explanation, at whatever price, save that of renouncing the sacred character of the two principles of thermodynamics.

When all of his attempts at deriving his conjecture had failed, he had recourse to what he called an act of despair: while still claiming to remain within the framework of the physical laws of the nineteenth century, he made a special reinterpretation of some of the rules of statistical physics so as to obtain the "good result," a result equivalent to his conjectured law for black bodies. We shall return below to the interpretation used by Planck, and to the opinion which the young Einstein held of it.

Disorder and Counting the Configurations of Fleas

We now return to the contents of Einstein's article from March 1905. This article contains several separate sections treating different aspects of light and its interaction with matter. The first section is an implicit critique of the work of Planck of which we have just spoken. In fact, Einstein shows that a correct application of the laws of physics, as known at the time, necessarily leads to a completely precise law for black-body radiation. Nevertheless, this law[4] had two faults: (i) it was in severe disagreement with experimental measurements in the range of large frequencies where Wien's law applies, and (ii) this law was physically absurd, since it predicts that a hot oven, or a simple wood fire, must emit an infinite amount of radiation, distributed predominately at very high frequencies. In other words, according to nineteenth-century physics, if one sat in front of a campfire, one would be instantly toasted to death, whatever the temperature of the fire might be! Einstein, of course, concluded from this first result that Planck's work, which claimed to derive a different law for black-body radiation while remaining within the framework of nineteenth-century physics, was neither mathematically nor physically coherent. However, his article does not contain any explicit criticism of Planck's work. This may seem slightly puzzling, since Einstein's correspondence in this epoch shows that, full of youthful fire, he was quick to criticize his physicist colleagues, including some of the most well known. The moderation shown within Einstein's article probably shows the influence of his closest friend at the time, Michele Besso, the same friend who helped him find the central idea of the theory of relativity. Indeed, in a letter from 1928, Besso wrote to Einstein:

> On my side, I have been your public during the years 1904 and 1905; by helping you to edit your communications on the problem of quanta, I deprived you of some of your glory; but, on the other hand, I procured you a friend, Planck.

However this may be, this first result demolished Einstein's confidence in the black-body law proposed by Planck. Because of this, in the rest of the article he only uses the previous law due to Wien, since he knew that it was experimentally very well confirmed for large enough frequencies. Starting from Wien's law, and using the laws of thermodynamics, he calculates the entropy of radiation of a given frequency, f, contained in a given volume, V. Recall that the *entropy* of a physical system is a measure of the *disorder* reflected in the incomplete knowledge that we have about the system.

Let us give an example, to clarify the notion of entropy and its connection with the idea of disorder. Consider a checkerboard: in other words, a board with 64 squares, eight squares long by eight squares wide. At an initial instant, we place a certain number of fleas on *one* particular square of this checkerboard. After this initial instant we let the fleas move freely, jumping randomly in every direction. Our only requirement is that the borders of this checkerboard are raised, stopping the fleas from escaping. After some time, in which the fleas have hopped everywhere, the ensemble of fleas will be distributed in a nearly uniform fashion on all the squares of the checkerboard. This *final state* is clearly more disordered than the *initial state* where one knew that all the fleas were assembled on a single, specific square. We may go even further and quantify the growth in disorder, between the initial and final states, by counting the number of possible configurations of the system of fleas. In the final state, each flea may be found, with equal probability, on any one of the checkerboard's 64 squares. The number of possible (equally probable) states for *one* flea is thus 64. If one has *two* fleas (considered as independent and distinguishable), the number of possible configurations for the system becomes 64 times 64, or 64 squared (64^2). For *three* fleas, we would have 64 cubed (64^3) and, more generally, the number of (equally probable) possible configurations for a system of n fleas will be 64 to the power n (64^n). By comparison, in the initial state, we have specified that all the fleas were on the same, known square, and thus had only one specific configuration for the system of fleas.

To summarize, the essential lesson to retain from this exercise of counting the configurations of fleas is the following: when a certain number of fleas, let's say n, are permitted to occupy a surface 64 times larger than the surface where they were initially confined, the number of possible configurations for the system is multiplied by 64 to the power n (64^n). If we had considered a different *ratio of areas*, let us say an area 10 times larger than the initial area, the number of possible configurations would have been multiplied by 10 to the power n (10^n). And if one had considered not fleas on a surface, but flies initially confined to a small volume, then released into the entire volume of a room, the number of possible configurations would have been multiplied, between the initial and final states, by the factor r^n, where r denotes the *ratio* between the final volume and the initial volume, and n the number of flies. The essential point for the following is that the number n of "independent corpuscles" in the system appears as the *exponent* of the ratio between the initial and final volumes available to the system.

Entropy and Disorder

In physics, given a system which is only specified by certain global, macroscopic characteristics, like its total energy or the volume in which it is contained, *entropy* is the logarithm of the number of possible microscopic configurations of the system (also called microscopic states). We recall that the logarithm of a number is, essentially, the number of digits (before the decimal, and not counting the first digit) of its decimal representation.[5] For example, the logarithm of 10 is 1, that of 100 is 2, and that of one million is 6. Note also that the logarithm of 1 is zero. In other words, the logarithm L of a given number N is such that $N = 10^L$. The notion of entropy had been introduced in the middle of the nineteenth century by Rudolf Clausius, who sought to better understand some fundamental work by the Frenchman Sadi Carnot. Clausius showed how to define the entropy of a system from other thermodynamic properties of the system and proposed, as an axiom, the famous second fundamental law of thermodynamics, according to which *the entropy of an isolated system can only grow*. (We recall that the first fundamental law of thermodynamics affirms that *the energy is conserved*.) Some years later, the Viennese physicist Ludwig Boltzmann understood that the origin of this second law was statistical,[6] and that the entropy was proportional to the logarithm of the possible number of microscopic states for the system considered.[7] This permitted the second law to be understood as a codification of the natural tendency of an isolated system to tend towards disorder. An example is furnished by the system of fleas considered above which, starting from an ordered initial state, will spontaneously evolve such as to successively occupy all possible states, and thus find itself, for most of the time, in a generic state, having lost the initial order.

The Neglected Equation: $E = hf$

In 1905, Einstein was one of the few physicists to understand in depth the connection between entropy and the number of microscopic states.[8] He thus knew how, starting from Wien's law for radiation in an oven's interior, to calculate the entropy, and thus the number of possible microscopic states for the radiation (for a fixed frequency f) contained within an oven of given volume. From this he derived the factor by which the number of possible microscopic states was multiplied when one increased the system's available volume by a ratio r. He found that this multiplicative factor for the number of states (or, equivalently, the factor by which the microscopic disorder gets multiplied) had exactly the form that we have obtained above in the case of fleas or of flies expanding into a greater volume: it took the form of a certain power of the ratio r, let's say r^n. Einstein concluded that the exponent n which appeared in this multiplicative factor could be interpreted as the number of independent light corpuscles present in the radiation of frequency f contained in the oven. Such was his principal argument for proposing the revolutionary idea that light had, like a bushel of fleas, a *discontinuous* (also called *discrete*) structure; that it was made up of separate *luminous grains*, which he called quanta of light.

Moreover, his reasoning gave a precise value for the exponent n, the number of independent light quanta present in the radiation. He found that this number n was obtained by dividing the total energy of the radiation (at the frequency f) by an expression of the type hf, where h was a universal constant, and f the frequency of the radiation. From this Einstein concluded that everything behaved as if light was made up of grains, with each grain of light having an energy E proportional to the frequency f of the light: $E = hf$. The universal constant h takes the value $6.626 \times 10^{-27} \mathrm{g} \cdot \mathrm{cm}^2 \cdot \mathrm{s}^{-1}$ and is called Planck's constant.[9]

The equation $E = hf$, obtained by Einstein in March 1905, is even more important and fundamental than the equation $E = mc^2$ that he would obtain in September of the same year. Nevertheless, the former equation is practically unknown, and the latter world-famous. Let us clarify that Max Planck was the first, in December 1900, to associate to the frequency f of black-body radiation the quantity of energy $E = hf$. However, the equation $E = hf$ only began to acquire all of its physical meaning in Einstein's articles from 1905 and 1906. For this reason, the equation is often referred to as the Planck-Einstein equation.

Contrary to what the usual popular treatments may lead one to think, Planck never proposed, in 1900, that the energy of the matter making up the interior surface of the oven (and even less the energy of light) was "quantized" in units of $E = hf$, meaning that it could only take the values 0, hf, $2hf$, $3hf$, $4hf$, etc. Planck introduced what he called energy elements $E = hf$ as an artificial tool in the calculation, to give meaning to the "number of microscopic states" of the matter lining the interior of the oven. Roughly speaking, he used these energy elements just as we previously used a checkerboard (with a finite number of squares) to calculate the number of configurations of fleas on a surface. In the calculation we made above, the result depended only on the *ratio r* between the final and initial surface areas. The size of an elementary square of the checkerboard disappears from the final result.[10] Planck, however, realized that the result of his calculation depended on the size of the elementary unit of energy, $E = hf$, which he used. He hoped that it would be possible, in the future, to attribute some meaning to his calculation while remaining within the framework of the physics of his time, that is to say with the energy of matter taking all the values from zero to infinity, and with light described by a continuous wave.

Einstein was the first[11] to think about the *physical discontinuities* (today called *quantum discreteness*) associated to the universal constant h. If Planck was the discoverer (in 1900) of the existence of a new universal constant of physics (and he understood from the beginning that this discovery represented the dawn of a new era in physics), Einstein (between 1905 and 1907) was the initiator of the physics of quantum discontinuities (and he understood how very revolutionary such a physics was).

First Consequences of the Neglected Equation

Having obtained his fundamental result that light was composed of grains of energy taking the value $E = hf$, Einstein finished his article of March 1905 by deriving several observationally verifiable consequences. Among these consequences, the most celebrated (since it would lead, after its experimental verification, to Einstein's Nobel prize of 1921) concerns the photoelectric effect. We have explained above why it seemed so mysterious to Philipp Lenard that a very great intensity of light was incapable of ejecting electrons from the surface of a solid, once the frequency f of the light wave became smaller than a certain threshold. On the other hand, Einstein's hypothesis of light quanta explains it easily.

Indeed, if we suppose that the basic process which leads to the extraction of electrons from a surface is that of a quantum of light with $E = hf$ giving its energy to an electron, the law of energy conservation allows one to conclude that the energy of motion (or kinetic energy) of the electron, after its exit from the solid, is $E_{kin} = hf - W$, where W is the energy (that is, the work) required to free the electron from the solid. This quite simple mathematical formula shows (since the kinetic energy must be positive) that it is only possible to extract electrons if the frequency of the light is greater than the threshold $f_{thresh} = W/h$. Moreover, when the frequency f is greater than f_{thresh}, this formula predicts a very simple relation between the kinetic energy of the electrons and the frequency of the light: $E_{kin} = h(f - f_{thresh})$. This prediction was quite precise, since Einstein also predicted the numerical value of the coefficient of proportionality h appearing in this formula, and the fact that the coefficient was *universal*, which is to say completely independent of the solid body from which one extracted the electrons. It took a dozen years of experimental work to verify Einstein's prediction in detail. The most precise verifications are due to the American physicist Robert Andrews Millikan in 1915. It is interesting to quote what Millikan himself said (in 1948) of his results:

> I spent ten years of my life testing that 1905 equation of Einstein's and contrary to all my expectations, I was compelled in 1915 to assert its unambiguous verification in spite of its unreasonableness, since it seemed to violate everything we knew about the interference of light.

We have already cited Max Planck's quote (dating from 1913) where he said that Einstein had "missed the goal" with his hypothesis of light quanta. Let us also point out that as late as January 1924, Niels Bohr, Hendrik Anton Kramers, and John Clark Slater were in doubt of Einstein's concept of the quantum of light. All of this shows convincingly the "very revolutionary" character of Einstein's article of March 1905. But the young employee of the Patent Office did not stop there. During the years between 1905 and 1924, he continued to explore the quantum discontinuities and their physical consequences. Let us briefly indicate some of the most important results obtained by Einstein.

Matter and Quanta

Bern, Switzerland, March 1906

Contrary to what is often written, the first physicist to claim that the energy of matter was physically quantized, (meaning that it could only take certain discontinuous values) was Einstein in March 1906, and not Planck in 1900 (see the previous discussion of what Planck did in 1900). In a continuation of the article discussed in detail above, Einstein returned in March 1906 to the contradiction which he had pointed out between the physics assumed at the time and the experimentally observed law of black-body radiation. He showed that the black-body law proposed by Planck in 1900, which was in perfect agreement with the measurements of the Berlin experimentalists, could only be derived from the general laws of statistical physics (by the counting of possible microscopic states) with the hypothesis that the energy of each "material oscillator" present in the interior surface of the oven only took a discontinuous series of values: 0, hf, $2hf$, $3hf$, etc. Here, as in Planck's work, the atoms present in the interior surface of a hot oven, and responsible for the absorption and emission of the thermal black-body radiation, are modeled as electric charges attached to a spring and oscillating around their equilibrium position. The quantity f denotes the frequency of oscillation of this spring. One must suppose that the surface is made up of an infinite number of oscillators covering all possible frequencies, since each oscillator (that is, each charge attached to a particular spring) will only be sensitive to light having the same frequency f as the oscillator considered.

Perhaps it has been noticed that the equation proposed by Einstein in 1906 for the *quantization* of the energy of matter, $E = nhf$, where n is a whole number ($n = 0, 1, 2, 3, \dots$), strongly resembles the relation which he had proposed the previous year for the energies of quanta of light. Its physical meaning is nevertheless different (and differs as well from the use of elements of energy by Planck in 1900). Here, E denotes the energy of a material system (a mass attached to a spring), and f the frequency of oscillation of this mass. In complete contradiction with Newton's laws of mechanics (and their relativistic modification given by the theory of relativity) which say that a mass attached to a spring can oscillate with any amplitude whatsoever, and thus with any energy whatsoever, Einstein dared to assert that the energy of oscillation of the spring can only take a discontinuous series of values 0, hf, $2hf$, etc., without being able to take intermediate values. Although this assertion was just as revolutionary as his hypothesis concerning the quanta of light, it was accepted more rapidly by physicists. The first to accept it was probably Planck in 1908. It must be said that this hypothesis, even if it contradicted Newton's mechanics, offered no major contradiction with well-known experimental facts. The constant h was numerically so small that it was impossible to see, in the laboratory, the quantization of the energy of a usual, macroscopic mechanical oscillator. On the other hand, the hypothesis concerning quanta of light seemed to be in violent contradiction with the numerous experimental verifications of the wave-like nature of light (even if Einstein had remarked, in 1905, that optical observations in the laboratory only tested average values and thus could possibly be explained just as well by light of a grainy nature).

The Icy Diamond

Bern, Switzerland, November 1906

At the end of 1906, Einstein was still working at the Patent Office, and therefore still had very little free time to devote to physics. However, he had a decisive advantage over most other physicists: he was the only one to truly believe in the quantization of the energy of matter. This left him the time to reflect calmly on the possible consequences of this quantization, in other systems than that of black-body radiation. He perhaps remembered the mysterious experimental results which had been obtained by his physics professor at the Zurich Polytechnic, among others. Heinrich Weber had studied the *specific heat* of certain solid materials, notably diamond. The *heat capacity* of an object is the amount of heat that must be supplied to it to raise its temperature by one degree Celsius. Of course, this quantity is proportional to the mass of the object considered. It is thus useful to consider the heat capacity of a sample from some substance which contains a fixed number of atoms (let's say 6.022×10^{23}, which is called Avogadro's number). We shall call this latter quantity the specific heat of the solid considered.

In 1819, the French physicists Pierre Dulong and Alexis Petit made a striking discovery. They found that the specific heat of a great number of simple substances was the same, taking the value of around six calories per degree (and per mole). The remarkable universality of the specific heat for (simple) solids was explained theoretically, 50 years later, by Ludwig Boltzmann. Boltzmann's explanation relied on recent results concerning the statistical interpretation of heat. Essentially, Boltzmann interpreted the heat of a solid as the energy of vibration of each atom around its equilibrium position. In 1876, through calculations using the statistical theory which he had just founded, he derived the ratio between this energy of vibration and the temperature of the object. He found exactly the same result as Dulong and Petit.[12]

Unfortunately, in 1876, it was found that some solids have a specific heat which is much smaller than six calories per degree. This is the case notably for boron, silicon, and diamond (or graphite, which is made, like diamond, from carbon atoms). In 1875, Weber greatly clarified the problem by showing that these three exceptions returned to ranks at higher temperatures. He found experimentally that the specific heat depends on the temperature of the object considered, and (generally) tends towards the simple, universal value of Dulong and Petit when the temperature grows very high. It nevertheless remained quite mysterious why, when the temperature is lowered, the specific heat of these three solids becomes much smaller than the value of six calories per degree, which was the prediction made by classical statistical physics. (We here use the adjective *classical* to characterize physics before the *quantum* revolution.) Even more so in that, in the case of diamond, its specific heat is only on the order of 1.4 at usual ambient temperatures. Thus, this unusually low specific heat is not some new phenomenon which is only apparent at very small temperature.

Einstein understood, in November 1906, that the idea which he had proposed in March of a quantization of the vibrational energies of mechanical oscillators furnished the key to the problem. The essential physical reason is the following: according to Boltzmann, the heat in a solid is stored in the form of the vibrational energy of its atoms, moving around their equilibrium positions. The precise calculation performed by Boltzmann had supposed that this vibrational energy could vary in a continuous fashion between zero and infinity. From this, he found that the vibrational energy stored in heat at an absolute temperature T was proportional to T. Einstein undertook a similar calculation, but now imposed that the vibrational energy of each atom, oscillating at frequency f around its equilibrium position, could only take the quantized values 0, hf, $2hf$, $3hf$, etc. He then found that the specific heat was zero at very low temperature, and grew slowly with temperature to attain the value predicted by Boltzmann when the atomic energy of thermal agitation became much larger than $E = hf$. One can intuitively understand

Einstein's result by thinking of the ambient heat as an agitating force, and of each atom as a child on a swing. If the amplitude of oscillation of the swing cannot grow continuously from zero, but must jump from a zero amplitude to the first amplitude of nonzero excitation, then to a second amplitude at a higher level, etc., an agitating force that is too weak to cause the first jump will leave the child in the state of lowest energy, with an agitation energy of zero.[13]

Einstein then compared his theoretical prediction to the experimental results of Weber and others, and found that the simple mathematical formula which he had found for the specific heat of a solid could account in a remarkable way for the observed data.[14] The fact that diamond has unusual behavior, of quantum origin, at room temperature is ultimately connected to the fact that diamond is very durable. Dear reader, I hope that from now on when you shall touch with your finger, or place on your skin a diamond, and feel the heat which it takes to warm itself, you shall remember that this everyday phenomenon confirms the idea, proposed by Einstein in 1906, of the quantization of the vibrational energy of material oscillators!

The Idea behind the Laser

From 1905 to 1911, Einstein obtained spectacular results on the frontier of several independent lines of research: relativity, the agitating motion caused by heat (Brownian motion), quanta, and the generalization of relativity in the presence of gravitation. However, from 1911 to 1916, he concentrated his mental energy almost exclusively on what was going to become the theory of general relativity. Independently of the fact that he understood (around 1911) that his principle of equivalence (see Chapter 3) offered him a key to opening a new domain of physics, he had spent so much time trying without success to understand quanta that he was happy to not think of it for some time. One can sense his frustration *vis-à-vis* the problem of obtaining a rational understanding of quantum discontinuities in what he writes to Michele Besso in May 1911:

> I no longer ask myself if these quanta really exist. All the same, I am no longer looking to construct them, for I know now that my brain is unable to make progress in this way. But I do very carefully explore their consequences, in order to learn what is the proper domain of application of this idea.

The problem which Einstein was confronting, along with everyone else who was then interested in quanta, was the stubborn presence of logical contradictions between the various hypotheses that it was necessary to make, in order to explain the ensemble of observed facts. For example, experiments on the interference of light were explained by describing light as a wave, while the photoelectric effect was understood, since Einstein's work, by supposing that light was a collection of quasi-independent corpuscles. Naively, one could have hoped (as Planck and Lorentz hoped) that light was really a wave, and that the corpuscular features were only some sort of artifacts arising from the interaction between light and matter. However, through a profound application of the link between entropy and probability, Einstein showed in 1909 that the fluctuations of energy contained in black-body radiation inside a small volume were equal to the sum of two different terms: one of these terms was readily understood if light was a random superposition of continuous waves, while the other could only be easily understood if light was an ensemble of independent, point-like corpuscles. Einstein then wrote:

> I think that the next phase in the development of theoretical physics will bring us a theory of light which can be interpreted as a sort of fusion between the wave theory and the theory of emission [of corpuscles].

In spite of his relentless efforts between 1909 and 1911, Einstein did not find a clear theoretical framework describing such a fusion between the wavelike and corpuscular aspects of light, from whence came the frustration which he expressed in 1911 to Michele Besso.

In 1916, in order to rest from the immense effort (ultimately crowned with success) which he had just dedicated to general relativity, he returned to his "very revolutionary idea" of the quanta of light and obtained several results of fundamental importance for twentieth-century physics. His point of departure consisted in combining the idea of light quanta with the idea, proposed by Niels Bohr, of the quantization of the energy of an atom.

In 1913, Niels Bohr had generalized Einstein's considerations (from 1906) claiming that the possible energies of an oscillator, such as a mass connected to a spring, are only 0, hf, $2hf$, etc., where f is the frequency of oscillation. In an act of great intellectual audacity, he had postulated that the possible energies for an arbitrary atomic system could only take a discontinuous series of values: E_0, E_1, E_2, etc. He had then assumed that the light emitted by an atom only contained certain frequencies (called transition frequencies), calculated from the differences between two of the possible energies through a generalization of the Planck-Einstein equation: for example, $hf_{10} = E_1 - E_0$ for the radiation associated to a quantum transition where an atom passes from an initial state with an energy E_1 to a final state with energy E_0. Finally, and this was the most innovative aspect of Bohr's work, he had postulated a new principle to determine the discontinuous list of possible energies. In the simplest case of a hydrogen atom (an electron in orbit around a proton), his new principle consisted of demanding that the *action* of the orbital motion of the electron—which is to say, for a circular orbit, the product of the "quantity of motion" (or momentum) $p = mv$ of the electron by the perimeter of its circular orbit—is equal to a whole multiple of the famous Planck constant, h.

In Einstein's work from 1916, he considers the following situation: he places a collection of atoms (having only a discontinuous series of possible energies E_0, E_1, E_2, ...) inside an oven which is heated to a certain temperature. We recall that such an oven produces, within it, a radiation whose energy is distributed over all frequencies. This distribution of frequencies is precisely what is called the law of black-body radiation. At this stage, Einstein does not assume that this law is known. He then writes that the entire system under consideration (oven, radiation, and the gas of atoms within the oven) reaches what is called thermodynamic equilibrium: an equilibrium state where, on average, each part of the system contains a constant energy, although it exchanges energy at every moment with the other parts. Without entering into the details, let us simply state Einstein's new results:[15] (i) a proof that the quanta of light emitted or absorbed by the atoms, during a quantum transition between two possible energies (let's say E_0 and E_1), carry a momentum $p = hf/c$, where f is the *transition frequency* associated to the difference between the two possible energies considered, and c is the speed of light, and (ii) the discovery of a new quantum process: the illumination of an atom by incident radiation of frequency f *stimulates* the atom to undergo a transition from the state of higher energy (E_1) towards the state of lower energy (E_0), by *emitting* a light quantum of energy hf and momentum hf/c moving in the same direction as the incident radiation. (During this process of *stimulated emission*, the atom "recoils" with the opposite momentum.)

These two results have had great importance in the development of twentieth-century physics, the former because it truly represents the first, purely theoretical discovery of a new *elementary particle*: the photon,[16] a particle of light possessing both an energy $E = hf$ and a momentum $p = hf/c$. This concept was a significant enrichment of the concept of quantum of light, introduced 11 years earlier by Einstein himself.

The latter result, that of *directed, stimulated emission* of light from atoms illuminated by incident radiation, is the basic concept which is put to use in the laser, whose importance is well known in both fundamental research and modern technology. This is not to say that Einstein invented the concept behind the laser in 1916. No, this would deny the importance of the numerous conceptual and experimental advances (due notably to the Americans Charles Townes and Arthur Schawlow, the Russians Nikolai Basov and Alexander Prokhorov, and the Frenchmen Alfred Kastler[17] and Jean Brossel) which led to the laser and to modern quantum optics. However, it does help us to appreciate the continuity of the chain of knowledge, from a purely theoretical investigation concerning the reality of the quanta of light, to the applications which support our daily life.

Light and Matter

Towards the end of June 1924, Einstein received a letter from a young Indian physicist, Satyendra Nath Bose. Although the name Bose was unknown to him, Einstein immediately became deeply interested in the new idea proposed by Bose, and applied by him in an article attached to the letter. In this article, Bose proposed to take seriously the idea (due to Einstein) that light consisted of luminous corpuscles, carrying an energy and a momentum. More precisely, Bose described black-body radiation, the light contained within a box heated to a certain temperature, as a *gas of light quanta*. He then applied the statistical methods of Boltzmann,[18] modified by the quantum considerations of Planck and Einstein, to the calculation of the thermodynamic properties of such a gas, and thus arrived at a new derivation of the law governing the frequency distribution of black-body radiation. Recall that this law, after having been "guessed" by Planck in October 1900, had already been the subject of several theoretical attempts to justify it: by Planck in December 1900, then by Einstein in 1906 and, in greater generality, 1916. Although Bose's new derivation was itself still imperfect, it had the great advantage, over previous derivations, of not appealing to either Maxwell's theory of electromagnetic waves, or to the properties of interaction between matter and light. Bose concentrated uniquely on the statistical properties of a gas made of a collection of luminous quanta moving inside a box.

After translating Bose's article into German, and sending it for publication to a German scientific review, Einstein understood that the new idea contained in this article could be greatly generalized. Roughly, Einstein's reasoning was the following. Before 1905, everyone believed that light and matter were fundamentally different: light was described as a continuous wave, while matter was described as a discontinuous collection of point-like particles. Between 1905 and 1916, Einstein had shown that it was fruitful to describe light as a discontinuous collection of luminous corpuscles, that is to say, essentially, as a gas of particles. Bose had just demonstrated that by applying quantum ideas (due to Planck and Einstein) to the statistics of such a gas, one recovered properties that had previously been obtained by appealing to the wave-like properties of light. All of this showed that there was, when one used quantum ideas, a deep similarity between light and matter. From this, Einstein had the idea to generalize the statistical calculation of Bose to the study of the thermodynamic properties of a gas of matter particles, within the framework of quantum ideas.

In the course of his work, Einstein obtained several new fundamental results, which have had a great impact on the development of twentieth-century physics. We note only two of these innovative results. The first is the discovery of a new physical phenomenon, of purely quantum origin, that is generally called *Bose-Einstein condensation*, although it was discovered by Einstein alone, in the articles which he wrote after reading Bose's work. This phenomenon of quantum condensation signals that a large fraction of the total number of particles, within a gas of material particles, may (when the density is large enough) be found in the quantum state of lowest energy, which classically would be the state of a particle at rest (without kinetic energy). This quantum condensation plays an important role in modern physics, both because it is the basic mechanism at work in remarkable phenomena such as superfluidity and superconductivity, and also because its very recent realization in the physics of ultra-cold atomic (or molecular) gases opens up the possibility for remarkable scientific and technological developments. We note in passing that, as had been the case for several of Einstein's revolutionary discoveries, this result was rather coldly welcomed by his colleagues, who doubted the possibility of this quantum condensation. The physical relevance of this phenomenon was only understood about 15 years after its prediction by Einstein.

The second of Einstein's innovative results, in his work from 1924, concerns the similarity between light and matter. Using one of his favorite theoretical tools, the treatment of thermodynamic fluctuations, Einstein studied the fluctuations in the number of (quantum) material particles contained within a subvolume of the total volume of the gas considered. This was a generalization of what he had studied in the case of fluctuations of black-body radiation. He found that these fluctuations were given by the sum[19] of two terms. The first of these terms was well known and represented the usual statistical fluctuations that were expected for a number of classical (that is, treated without quantum considerations) particles. The second term was of a different sort, and Einstein proposed its interpretation "as a term accounting for the fluctuations caused by the interference of radiation processes properly associated to the gas." In other words, by a reversal of the attribution of corpuscular properties to light (well known for its wave-like properties) through the introduction of quanta of light, Einstein here proposes to attribute wave-like properties to material particles. It seems that Einstein came to this conclusion in September of 1924.[20] At the end of November, or the beginning of December, 1924, he read the doctoral thesis of Louis de Broglie, sent to him by Paul Langevin, and understood that de Broglie had already (in 1923) proposed to associate a wave field to a material particle.

Because of this, in his second article on the quantum theory of an ideal gas, finished in December 1924, Einstein, after having indicated the thermodynamic reasoning which suggests the necessity to associate a radiation process to a gas of material particles, explicitly mentions de Broglie's "very remarkable" thesis and delineates the undulatory characteristics associated to a material particle. These are (a posteriori) natural generalizations of Einstein's results concerning the quanta of light. To a wave of light of frequency f, and thus of period $T = 1/f$ and wavelength (denoted by the Greek letter lambda) $\lambda = cT$, Einstein had associated a particle of energy $E = hf = h/T$ and of momentum $p = hf/c = h/\lambda$. As for de Broglie, he proposed to associate with each particle of energy E and momentum p a wave of period T and wavelength λ, such that one still had[21] $E = h/T$ and $p = h/\lambda$.

With this work, completed in December 1924, the list of Einstein's truly path-breaking contributions—which had opened the way to the very essence of twentieth-century physics by founding those three great scientific theories which are special relativity, general relativity, and quantum theory—comes to an end. Einstein still contributed other important scientific ideas (some of which have only recently flowered, and others which are still germinating), but we are forced to recognize that from 1925 on, he ceased to be the leading figure in theoretical physics the whole world over. We shall nevertheless be able to see the more subterranean importance taken by some of the research which he pursued in the later part of his life.

Einstein in his study at Princeton, 1954. (*Credit Rue des Archives / TAL.*)

6

Confronting the Sphinx

I rejoice only moderately in the great recent discoveries, since I do not find
for the moment that they help me to understand the fundamental issues.
Also, I feel like a kid who does not succeed in learning the alphabet,
although, as strange as it may seem, I have still not lost hope. But after all,
it is indeed the sphinx who we are dealing with, not a willing streetwalker!
—Einstein, letter to Max von Laue, March 23, 1934

A Crucial Conversation

Berlin, Germany, early 1926

The young Werner Heisenberg was awed and impressed on entering the
physics seminar room of the University of Berlin on this day early in 1926.[1]
He was only 24 years old, and had been invited to give a lecture on the
"new" quantum mechanics, which had just been born. While rather fever-
ishly throwing a final glance at his notes, he saw, taking seats in the front
row, the upper crust of the international physics community: Max Planck,
Walther Nernst, Max von Laue, and others. The faces of these physicists,
famous for their fundamental discoveries, held all of the seriousness and rigorous composure of German academic life. Then, just before the hour set
for the beginning of the lecture, the physicist who impressed him most, he
whose work he had admired since adolescence, when he had discovered the
theory of general relativity in a book[2] entitled *Space, Time, Matter*, he whose
letters were read aloud by his professor and thesis advisor in Munich, Arnold
Sommerfeld, to illustrate his course: Albert Einstein entered the room and sat
down in the front row, giving him a friendly smile, partly to excuse himself
for nearly arriving late, and above all to put him at ease.

Thus given confidence, Heisenberg began to relate the physical concepts and mathematical formalism of the *new* quantum theory. Indeed, in the last few months there had developed, with unheard-of speed, a new mathematical formalism which was hoped to supplant the "old" theory of quanta. The old theory of quanta was that disparate collection of mutually contradictory ideas developed between 1900 and 1924, which attempted to describe the quantum discontinuities whose existence had slowly been revealed through the understanding of various physical phenomena. The discovery which had initiated the theory of quanta (the precise structure of black-body radiation) had been made here in Berlin itself, through the extremely precise measurements of Otto Lummer, Ernst Pringsheim, Heinrich Rubens, and Ferdinand Kurlbaum, and through Max Planck's theoretical "act of despair." But it was, above all, Einstein's collective work on quanta, between 1905 and December 1924, which had shown the need for a profound readjustment of physics (to which were added, starting in 1913, the innovative concepts of Niels Bohr who had shown how to apply the quantum ideas to atomic physics). The new quantum formalism which Heisenberg spoke of had come from some of Bohr's ideas on atomic structure, and some concepts introduced by Einstein in 1916 concerning the interaction between an atom and electromagnetic radiation. Einstein had introduced, among other things, some coefficients (denoted A), which measured the probability (per unit time) with which an atom, initially found in a certain (quantized) state, could experience a quantum transition towards another quantized state with lower energy by emitting, at a random instant and in a random direction, a quantum of light.[3] Heisenberg had been initiated into the physics of these quantum transitions by his thesis advisor in Munich, Arnold Sommerfeld, and by Max Born, at Göttingen. After having completed his thesis at the age of 22, he became Born's assistant at Göttingen in October 1923. In 1923 and 1924, Heisenberg worked under Born's direction, and learned from him several crucial ideas and techniques, notably the idea to introduce new coefficients, denoted a, associated like Einstein's coefficients A to the quantum transition between two states of an atom. Roughly speaking, the new coefficients a, called amplitudes of quantum transition,[4] were such that their squares were equal to Einstein's coefficients A.

The essential idea at the base of the new quantum theory had come to Heisenberg early in June 1925, while he was recovering from a bad bout of hay fever by spending some time on the island of Heligoland, to the north of Germany. This idea consisted in replacing the usual notion of a continuous orbit describing the possible motion of an electron[5] around an atom by the collection of amplitudes a, associated to the transitions between the atom's possible quantized states. Each transition amplitude is defined by supplying two numbers: the number fixing the initial energy state within the discontinuous list of possible quantum states of the atom, and the number fixing the final state. The total collection of amplitudes is thus analogous to a checkerboard or a multiplication table,[6] of which each elementary square is fixed by supplying two numbers: one number fixing the horizontal projection of the square in question, the other fixing its vertical projection.

While Heisenberg was explaining the motivations which had led him to replace the description of the continuous orbit of an electron in an atom by such checkerboards of transition amplitudes, he looked with worry out of the corner of his eye to where Einstein was seated, to see how he was reacting to the introduction of such "witches' multiplication tables."[7] While not convincing him, Heisenberg succeeded in drawing Einstein's interest, particularly when, at the end of his lecture, he indicated that the new rules of multiplication for two amplitude tables, introduced by him and developed in recent work done in collaboration with Max Born and Pascual Jordan, permitted one to demonstrate, through detailed calculation, Einstein's result which said that the energy fluctuations of the radiation contained within a subvolume were the sum of two separate terms: a term connected to the undulatory character of the radiation and a term connected to its corpuscular character.[8] This result, concluded Heisenberg, showed that the new quantum formalism was capable of describing the undulatory and corpuscular aspects of a continuous field (such as the electromagnetic field) *at the same time.*

After the colloquium, Einstein came to congratulate Heisenberg on his remarkable results, and asked Heisenberg to accompany him home in order to discuss in more detail the new ideas at the base of the formalism which he had presented. Upon arriving at his apartment, Einstein asked him to again explain the physical motivation leading to the replacement of the notion of a continuous orbit by that of an infinite table of transition amplitudes.

Let's listen to a central part of their dialogue, such as it was later reconstructed by Heisenberg himself:[9]

HEISENBERG: ... Since it is reasonable to allow into a theory only directly observable quantities, I thought it more natural to restrict myself to these frequencies and amplitudes,[10] bringing them in, as it were, as representatives of electronic orbits.

EINSTEIN: But all the same, you do not seriously believe that a physical theory should only include observable quantities?

HEISENBERG: I thought that it was you yourself who had made this idea the foundation of your theory of relativity. You stressed that one could not speak of an absolute time, since one cannot observe this absolute time. You said that only the readings of clocks, made in a system of reference either in motion or at rest, were able to determine the measurement of time.

EINSTEIN: Perhaps I used this sort of philosophy, but it is nonsense nevertheless. Maybe, to express myself more prudently, I will say that from a heuristic point of view, it could be useful to remember that which one really observes. But, at the level of principles, it is completely erroneous to want to found a theory uniquely on observable quantities. For, in reality, things happen in exactly the opposite way. *It is only the theory which decides what can be observed.*

We have emphasized the final sentence since it resonated for a long time in the young Heisenberg's mind, and played a crucial (and generally unknown) role in the later development of the quantum theory. Let us only say here that this message (it is the theory which decides what is observable) had been inculcated into Einstein by the years spent in the erratic construction of general relativity. For years, the connection (so clear in *special* relativity) between the coordinates of space and time and the measures of distance and duration had remained obscure in *general* relativity. Einstein had only worked his way free of confusion at the end of 1915 when he understood, after having constructed the theory, that it was the very mathematical formalism of general relativity which permitted one to define *a posteriori* that which was observable when space-time was deformed by matter.

"Waves over Here, Quanta over There!"

In the beginning of the year 1926, close to the time when Heisenberg had given his lecture at the Berlin colloquium, another mathematical formalism had been proposed, by the Austrian theoretical physicist Erwin Schrödinger, to supplant the "old" Planck-Einstein-Bohr theory of quanta. This formalism, called wave mechanics, had, according to Schrödinger himself, taken root in the ideas of Louis de Broglie, and in the "brief but infinitely clairvoyant" remarks made by Einstein (within letters, and in the article from the end of 1924 discussed in the preceding chapter). Schrödinger's *wave mechanics* seemed completely different from the Born-Heisenberg-Jordan *matrix mechanics*. In the former, the state of the system considered (let's say electrons orbiting around the nucleus of an atom) was described by a wave amplitude $\mathcal{A}$, which was a *continuous* function[11] of time and of the coordinates of the electrons, while the latter only considered the *discontinuous transitions* between the various possible stationary states of the atom, and described them by an infinite checkerboard of transition amplitudes a_{nm}. These two descriptions seemed to be antipodal to each other. The first gave a perfectly continuous image (in time, and in the space of configurations of the system), while the second was only interested in the discontinuous transitions experienced by the system. Nevertheless, Schrödinger quickly enough showed that there was a mathematical equivalence between the two formalisms. More precisely, he showed that knowledge of the wave equation describing the propagation of the continuous amplitude $\mathcal{A}$ permitted the simultaneous calculation of the possible stationary states of the system, their quantized energies, and the infinite checkerboard of transition amplitudes between these stationary states. Roughly speaking, the possible stationary states are analogous to the series of pure vibrational states of an elastic object, like those of a piano string which can vibrate in its fundamental mode, or in the mode corresponding to the first harmonic (one octave higher than the fundamental mode) or even in the second harmonic (a fifth above the first harmonic), etc.

In fact, it seemed for a long time that Schrödinger's wave description was more complete than the Born-Heisenberg-Jordan discontinuous description. Above all, Schrödinger's description seemed to suggest that one could perhaps even get rid of the idea of quantum discontinuity (despite all that it had allowed to be understood, including Einstein's theory of atomic transitions), and describe reality uniquely in terms of a continuous wave phenomenon.

Einstein had initially welcomed, with satisfaction and some relief, Schrödinger's formalism, which seemed to him closer to his deeply rooted intuition about reality than the "witches' multiplication tables" used by Heisenberg and companions. But he was rather rapidly disenchanted, first, because the wave amplitude $\mathcal{A}$ was not propagating in the usual three-dimensional space but in a space of six dimensions for a system of two particles, nine dimensions for a system of three particles, 12 dimensions for four, etc; and second, because wave mechanics had great difficulty in accounting for all of the experimental facts which had led Einstein and others, for around 20 years, to introduce the quantum discontinuities. In August 1926, Einstein summarized his sentiments in a letter to Paul Ehrenfest:

> Waves over here, quanta over there! The reality of each has the solidity of rock. But the devil makes them rhyme together (and the rhyme is well and truly real).

Einstein's dissatisfaction, on being confronted with the paradox that nature exhibits wave-like aspects and particle-like aspects at the same time, lasted until the end of his life. As we shall see, that which convinced most other scientists did not carry away Einstein's approval.

Einstein's Ghost Field, Born's Probability Amplitude, and Heisenberg's Uncertainty Relations

We shall not try to discuss, in even a slightly exhaustive way, the development of the physical interpretation of the mathematical formalism of quantum theory. We will only show the essential, though sometimes hidden, role played by some of Einstein's ideas.

The first crucial advance dates from the summer of 1926, and is due to Max Born. As he explicitly wrote:[12]

> I start from a remark by Einstein on the relation between [a] wave field and light-quanta. He [Einstein] said approximately that waves are there only to point out the path to the corpuscular light-quanta, and spoke in this sense of a "ghost field" which determines the probability for a light-quantum ... to take a definite path ...

These remarks by Einstein on a ghost field, or a pilot field, were communicated verbally by him to several scientists (Max Born, Eugene Wigner, and others) in the 1920s, but he never published them. However that may be, it seems that they motivated Born to propose the interpretation of the wave amplitude $\mathcal{A}(t, \mathbf{q})$ of a certain physical system as an amplitude of probability to find, at the instant t, the system in the configuration described by the variables $\mathbf{q}$. (As mentioned previously, when we consider a single particle, $\mathbf{q}$ denotes its three coordinates in space; but, when we consider a system of two particles, $\mathbf{q}$ denotes the six coordinates necessary to fix the spatial position of two particles; etc.) Born further explained (in a footnote added during proofreading) that the probability of finding a system in a configuration $\mathbf{q}$ was proportional to the square[13] of the amplitude $\mathcal{A}(\mathbf{q})$. Born then summarized the essence of the interpretation of quantum theory which he was proposing: "The motion of particles follows probability laws but the probability itself propagates according to a causal law."

The second part of Born's quote alludes to the fact that Schrödinger's wave equation, written by the latter in early 1926, is a deterministic equation of propagation, which determines in a unique way the temporal evolution of the amplitude $\mathcal{A}$, once one knows its value at an arbitrary initial instant.

Born's probabilistic interpretation was an important conceptual advance, but it raised more questions than it answered. In fact, it was a mere hypothesis, while it should have been derived from the mathematical formalism of the quantum theory. This is what Heisenberg believed during the end of 1926 and the beginning of 1927. Heisenberg was then working in Niels Bohr's group in Copenhagen. He held intense discussions with Bohr, which often lasted well past midnight, on the physical interpretation which should be given to the mathematical formalism of the quantum theory. In February 1927 Heisenberg, remaining alone in Copenhagen while Bohr was skiing in Norway, had a new idea destined to clarify the compatibility between a wave description and a corpuscular description for a single quantum particle (such as an electron). As he himself recalled,[14] the memory of his conversation with Einstein one year before played a crucial role in his thought process:

> That night, it was perhaps around midnight that I suddenly recalled my discussion with Einstein, and that I remembered his phrase: "Only the theory decides what one can observe." I realized immediately that it was within this remark that one must look for the key to the enigma which had so occupied [Bohr and me]. I then went for a nocturnal walk through the Fälledpark to reflect on the import of Einstein's comment.

It is in the course of this night-time walk, reflecting on the import of Einstein's phrase, that Heisenberg discovered his very famous uncertainty relations,[15] saying that the product of the "uncertainty" in the position of a particle and the "uncertainty" in its momentum[16] must necessarily be greater than Planck's constant h.[17]

Heisenberg understood that the uncertainty relations permitted a clarification of the conditions in which one could use the idea that a quantum particle is simultaneously described by a wave and by a corpuscle. For example, it seemed that the observation of rectilinear tracks, visible at the macroscopic level, left by particles in certain detectors implied that a particle must necessarily be described as a localized corpuscle. The uncertainty relations showed that the finite width of the tracks was compatible with a wave behavior on distance scales which were small compared with this width.

When Bohr returned from his vacation in Norway, Heisenberg enthusiastically explained to him what he had found by following Einstein's philosophy ("The theory alone decides what is observable"). In the interval, Bohr had continued his own reflections and had convinced himself that it was necessary to base the interpretation of quantum mechanics not on a logical derivation dictated by the theory itself (as Einstein had suggested) but on a new epistemological concept, introduced in ad hoc fashion for the interpretation of the quantum theory, called complementarity. As Heisenberg said, in Bohr's mind,

> complementarity should describe a situation where we could grasp a single and identical phenomenon by two different modes of interpretation [for example, wave and corpuscle]. These two modes must both mutually exclude and complete each other; and it is only the juxtaposition of these contradictory modes which allows one to completely exhaust the visual content of the phenomenon.

The discussion between the young Heisenberg (who was then 25 years old) and Bohr (whose 1913 work had played a crucial role in the development of the quantum theory) was rather stormy. Heisenberg admired Bohr as a scientist, and also venerated him like a father. He had expected that Bohr would appreciate the innovative conceptual advance represented by the discovery of the uncertainty relations. In place of this, Bohr expressed some reservations and offered some detailed technical criticisms. Above all, he only considered his own idea of complementarity to be general enough to serve as a basis for a coherent interpretation of the quantum theory. The tension between the two men was great, and led to permanent damage of their relationship. Confronted with Bohr's stubbornness, Heisenberg gave up on convincing him of the soundness of the general epistemological attitude suggested by Einstein, and reluctantly accepted the necessity of using an ad hoc interpretive language based on complementarity. Heisenberg published his discovery of the uncertainty relations, and their consequences for the interpretation of quantum reality, by himself, and left Bohr to prepare a detailed article on the idea of complementarity, which Bohr presented some months later at the Solvay council in the autumn of 1927.

A Watershed Moment

The fifth Solvay council, held in Brussels in the autumn of 1927, was a very important event. It was a watershed moment, both for the international community of theoretical physicists,[18] and for Einstein's scientific career. It is at this meeting that Einstein was first confronted with the interpretation of the new quantum theory proposed by Bohr, starting from ideas of Born (the probabilistic interpretation of the amplitude $\mathcal{A}$), and Heisenberg (the uncertainty relations), and from the concept of complementarity. Each of the theoretical physicists waited with a passionate interest to see Einstein's reaction. For everyone, Einstein was not only the greatest living physicist, but also the one whose revolutionary ideas had been crucial for the discovery and comprehension of quantum reality. The physicists of the younger generation (Heisenberg, Pascual Jordan, Wolfgang Pauli, etc.) worshipped Einstein, and considered themselves to be his modest successors. Was the pope of theoretical physics going to bless, on the baptismal font of complementarity, the new quantum child that everyone considered as his intellectual grandson? Well ... no! Einstein was not convinced by the interpretation of quantum theory defended by Bohr.

For many, the disappointment was great. And some (like Paul Ehrenfest) went so far as to compare Einstein's attitude *vis-à-vis* the new quantum mechanics to those of the opponents of the theory of relativity, who had been disconcerted by the novelty of Einstein's ideas and had refused to change their old habits. I think that the traditional image of Einstein as an aging revolutionary, refusing the new quantum ideas because they went against his prejudices about what reality must, *a priori*, be, is inexact. This does not mean that I think the attitude of Bohr, and of the majority of physicists who followed him by adopting what is called the Copenhagen interpretation of quantum theory, had been an error. Far from it! From a practical point of view, the consensus which crystallized at the Solvay council of 1927 around the Copenhagen interpretation helped the development of the new quantum ideas, and has permitted their application in an ever-growing domain of physics. A large part of the physics and technology of the twentieth century is based on the application of quantum theory (to the physics of solids, to atomic physics, to high-energy physics, etc.). The interpretive scheme proposed by Bohr at the 1927 Solvay council helped to put aside the obscure epistemological aspects of quantum theory, and enabled the exploration of the new world which was opened up by its mathematical formalism. However, having said that, I think that it is time (above all, on the occasion of the centenary of the revolutionary ideas proposed by Einstein in 1905) to give a description of Einstein's attitude *vis-à-vis* the quantum theory that is not a crude caricature, and at the same time to recognize both the fundamental soundness of his epistemological objections, and the visionary character of the works he undertook after 1927.

Fundamentally, I think that Einstein was not convinced by Bohr because the idea of complementarity was only a conceptually obscure and technically ill-defined cloak. In May 1928, in a letter to Schrödinger (who shared his doubts) Einstein compared the Copenhagen interpretation to a soft pillow, on which one could fall asleep without asking oneself questions about quantum reality:

> The tranquilizing philosophy (or, dare I say, the religion?) of Heisenberg-Bohr is so delicately put together that, for the moment, it furnishes to the true believer a soft pillow that he has a hard time leaving.

Later (in 1939), when Bohr had ossified into his posture as the apostle of complementarity, now a panacea for all of the problems of interpretation mentioned by Einstein, Schrödinger, and others, Einstein described Bohr (in a letter to Schrödinger) as a "mystic, forbidding any questioning about whatever might exist independently of the observer . . . ."

In a more precise fashion, I think that Einstein's dissatisfaction came from the fact that the Copenhagen interpretation was not in agreement with the idea which Einstein had expressed to Heisenberg, and which had led the latter to the discovery of the uncertainty relations: "It is the theory which decides what is observable." Bohr was adding an entire interpretive superstructure to the mathematical formalism of quantum theory, founded on the utilization of a special language, and having recourse to another scientific theory (classical Newtonian physics) which was supposed to apply to macroscopic objects (like the measurement instruments). It is because Einstein had very high standards of conceptual clarity that he could not be satisfied with "the tranquilizing philosophy (or religion?) of Heisenberg-Bohr." The clearest formulation that Einstein gave of his conceptual dissatisfaction is probably that which he expressed in 1932 in a letter to Wolfgang Pauli. We quote it such as it is, even if the Latin it uses is awkward: "Incidentally, I do not say *probabilitatem esse delendam*, but *probabilitatem esse deducendam*, which is not the same thing."

In other words, Einstein does not say that one must get rid of (*delendam*) the probabilities (which appear, according to Born, in the quantum theory), but that one must deduce (*deducendam*) the appearance of these probabilities (from the mathematical formalism which defines the quantum theory). Recall that Einstein was indeed an expert in the utilization of probabilities in classical physics (thermodynamics, Brownian motion), and it is he who introduced probabilities into quantum physics (in 1916, in his work on the absorption and emission of light by atoms). During the 20 or so years in which he had been (nearly) alone in believing in the quanta of light, he had spent countless hours trying to render the (deterministic) wave-like and (random) corpuscular descriptions of light compatible. He was not a man to resign himself to an abandonment of the logical, deductive character of science in favor of what the American physicist Bryce DeWitt recently called a "fuzzy metaphysics."

"The Marble Smile of Implacable Nature"

However that may be, starting in 1927, Einstein ceased following, in any detail, the advances made with quantum theory. On several occasions, he expressed his admiration for the remarkable results thus obtained, and his conviction that quantum theory represented "significant, and even, in a certain sense, definitive, progress in physical knowledge." But he hoped to be able to deduce the (quantum) probabilities from an underlying structure of reality. For 20 years, he had been the only one to believe in the quanta of light (even Bohr had doubted this concept until 1924). He had grown used to the solitary pursuit of his own path of research, even if most physicists thought that he was on a path leading nowhere. (This was what had happened to him during the years between 1907 and 1915 when he searched, in the dark, to generalize the theory of relativity.) He was thus not a man to take fright in being nearly alone in objecting to the Copenhagen interpretation of quantum mechanics.

In fact, some years before the watershed moment of the 1927 Solvay council, Einstein had begun to be interested in an ambitious program: that of finding a generalization of general relativity which could account, in a "unified" manner, for the existence of the electromagnetic field, along with the gravitational field. This program had been initiated by other scientists in the years between 1917 and 1921, notably by Hermann Weyl, Arthur Eddington, Rudolph Bach, and Theodor Kaluza. Starting in 1922, Einstein began to involve himself in this line of research. When he first engaged himself in this program, he did not know that he would devote the rest of his career to it, unfortunately without bringing it to a convincing result. Many times, he thought he had found a satisfactory geometric framework for generalizing general relativity, only to subsequently find an insurmountable physical fault in his trial theory. A passage written to Hermann Weyl in May 1923 about one such theoretical trial characterizes his state of mind well enough:

> The whole idea must be carried through and it is of a strange beauty; above it, however, hovers the marble smile of implacable Nature, which has given us more longing than intellect.

We shall not enter into the maze of research towards a "unified theory" which carried Einstein through the last 30 years of his life. This research kept him busy literally until his final breath, for on Sunday April 17, 1955, Einstein asked his secretary Helen Dukas to bring him, on his Princeton hospital bed, some writing material, as well as his most recent pages of calculation. In spite of the physical pain, Einstein absorbed himself on that Sunday, as he had done all his life, in the search for the rational structure of reality. He died a few hours later at 1:15 in the morning.

Let us note that his program of unification was supported by three distinct wishes. First, Einstein hoped to unify gravitation and electromagnetism through a new geometric structure of space-time. The idea was that a space-time "geometry" richer than the Riemannian geometry used in general relativity could naturally explain the existence and properties of the electromagnetic field, such as it had been described by Maxwell. Einstein's second hope was to unify the continuous and the discontinuous by describing a particle as a region of space where the continuous field which it creates is very intense, without, however, becoming infinitely large. Einstein had the idea that gravitation, that is to say the deformation of space-time, was essential to avoid what happened in Maxwell's theory in "flat" Minkowski space-time. In this latter case, an electrically charged, point-like particle creates an electric field which becomes infinitely large at the particle's position. The field is said to become singular, and the particle is thus a *singularity* of the field. The third hope of Einstein's program was to account for quantum phenomena in the framework of a classical (in the sense of nonquantum) field theory.

This very ambitious program was not completed by Einstein (however, see the concluding chapter of this book). Like a modern Sisyphus, he saw the successive theories which he had constructed, stone by stone, for years, topple down again and again. But this never discouraged him. Indeed, he perfectly illustrates both Albert Camus' thought that "one must imagine Sisyphus happy" and that of Gotthold Lessing saying that "the search for truth is more precious than its possession." Einstein was sustained in his quest for harmony by the profound sentiment (which he himself qualified as religious) of being a part of a vast harmonious reality which partially manifested itself to the human consciousness through the beauty and rationality of the universe. We can listen to him describing this sentiment in a passage written in 1930:

> The most beautiful thing we can experience is the mysterious. It is the fundamental emotion that stands at the cradle of true art and true science. He who does not know it and can no longer wonder, no longer feel amazement, is as good as dead, a snuffed-out candle. It was the experience of mystery—even if mixed with fear—that engendered religion. A knowledge of the existence of something we cannot penetrate, our perceptions of the profoundest reason and the most radiant beauty, which only in their most primitive forms are accessible to our minds—it is this knowledge and this emotion that constitute true religiosity; in this sense, and in this alone, I am a deeply religious man. I cannot conceive of a god who rewards and punishes his creatures, or has a will like our own. Neither can I, nor would I want to, conceive of an individual who survives his physical death; let feeble souls, from fear or absurd egoism, cherish such thoughts. It is enough for me to contemplate the mystery of conscious life perpetuating itself through all eternity, to reflect upon the marvelous structure of the universe which we can dimly perceive and to try humbly to comprehend even an infinitesimal part of the intelligence manifested in Nature.

Adventurers in Entangled Reality

Even if Einstein did not succeed in making good his very (too?) ambitious program of unification, several other works which he finished after 1925 have revealed themselves, in the long run, to be quite important. The unique greatness of Einstein is measured by the fact that any one of his works of secondary importance would have sufficed to assure the career of a scientist of ordinary stature. Before giving some succinct remarks concerning Einstein's later work, let us indicate where they were completed. Earlier, we left Einstein in Berlin, installed with the title of director of research within the Prussian Academy of Sciences. Einstein worked in Berlin from 1914 to 1933. During this period, he divorced his first wife, Mileva (who raised his two sons, Hans Albert and Eduard, in Zurich, with Einstein's financial assistance). In 1919, he married again, with his cousin Elsa Einstein Löwenthal. This marriage would last until Elsa's death in 1936, in Princeton. They had no children, but Elsa's daughters, Ilse and Margot, lived for a long time with the couple, and Margot continued to live with Einstein when he became a widower. In 1933, Hitler's rise to power and the mounting anti-Semitism in Berlin forced the Einstein family into exile. In France, Einstein was offered (through Paul Langevin's initiative) a chair at the Collège de France. After having initially accepted this offer, Einstein judged it safer to accept the offer made to him by the newly created Institute for Advanced Study in Princeton, New Jersey. From October 1933 until his death on April 18, 1955, Einstein worked at the Institute for Advanced Study. There he found a peaceful haven where he could concentrate on his work, in an isolation which grew throughout the years, separating him more and more from the majority of the scientific community.

A complete summary of Einstein's later significant works should mention: (i) his work (with Nathan Rosen) on the topological structure of a space deformed by the presence of one or several particles, and on the possibility that a particle does not constitute a singularity of the geometry, but rather a regular "bridge" connecting separate spaces (Einstein-Rosen bridges); (ii) his work on gravitational lensing, which is the augmentation of the apparent luminosity of a star whose light is deflected, before reaching the observer, by the gravitational field of a mass distribution situated near the line of sight; and (iii) his work (with Leopold Infeld and Banesh Hoffmann) on the motion of many bodies, represented as point-like singularities of the geometry.

Each of these works has initiated, after Einstein's death, an extremely rich research direction which has important repercussions in present-day science. For example, the study of gravitational lenses became important after the first observation, in 1979, of an effect of this type. The importance of this effect has only grown through the course of recent years, and should grow even more in the future. As for the Einstein-Infeld-Hoffmann work on the motion of several singularities of the geometry, their methodological importance was understood in the middle of the 1970s when modern astrophysical developments obliged theorists to consider the orbital motion of gravitationally condensed objects, like neutron stars or black holes.[19] The theoretical study of the motion of two black holes is of great present-day importance, and it attracts cutting-edge research, for it is an essential element in the determination of the behavior, and thence the possibility of detection, of the gravitational waves emitted from the coalescence of binary black hole systems.

But we shall here concentrate on another work from the Princeton phase of Einstein's career, that which he completed in 1935, in collaboration with Boris Podolsky and Nathan Rosen. This work illustrates well the visionary profundity of Einstein's approach towards physics. We have remarked previously on Einstein's refusal, in 1927, to accept the "soft pillow" of the Copenhagen interpretation of quantum theory. For several years, Einstein hoped to find a technical fault in this interpretation, for example, in the form of a subtle violation of the uncertainty relations. Rapidly enough he convinced himself of the absence of such faults. He then made an effort to more finely characterize his dissatisfaction *vis-à-vis* the Copenhagen interpretation, and his feeling that either this interpretation, or the quantum theory itself, was incomplete. The article by Einstein, Podolsky, and Rosen (EPR for short) marks a very important stage in the understanding of the deep structure of quantum theory. Indeed, this article brought to attention a paradoxical aspect of the formalism of quantum theory: the *entanglement* of two physical systems which have interacted (quantum mechanically) in the past, before separating.

Let us give an example of such an EPR situation. Consider a system of two particles. For simplicity, we shall suppose that the masses of the particles are equal to each other. Heisenberg's uncertainty relations say that one cannot measure, with infinite precision, both the position and speed of the first particle at the same time. Likewise, they forbid a precise simultaneous measurement of the position and speed of the second particle. Nevertheless, it can be shown that nothing forbids the specification (or measurement), with infinite precision, of both the position of the midpoint (the center of mass) between the two particles and their relative speed. Because of this, one may initially prepare the system of two particles in a quantum state where the midpoint between the two particles is a well-localized point, that we can take as the origin of coordinates, and where, moreover, the relative speed is zero. We let this system evolve freely from this initial state. Then, at a certain moment, we make observations (very far from the origin of coordinates) on one of the two particles, let's say the first. Heisenberg's relations forbid the *simultaneous* measurement of the position and speed of the first particle but nothing, in quantum mechanics, forbids the measurement, with infinite precision, of one or the other. Imagine first that we were measuring the position of the first particle and found it to be equal to a certain value x_1. As we know that the midpoint of the particles is fixed at the origin of coordinates, we deduce from this measurement that the position of the second particle is well determined, and takes the value $x_2 = -x_1$. However, imagine that we had decided to measure not the position of the first particle, but its speed, and that we had found a certain value v_1 for this speed. Since we know that the relative speed $(v_1 - v_2)$ between these particles is zero, we deduce from this measurement that the speed of the second particle is well determined, and takes the value $v_2 = v_1$.

Thus, according to the arbitrary choice that one makes on the fashion in which one observes the first particle, one can determine, with certainty, the position or the speed of the second particle without directly observing it and thus without disturbing it in any way. Einstein, Podolsky, and Rosen assumed that every certain prediction that one could make for a system, without perturbing it in any way, must correspond to something "real." They thus deduced from the thought experiment which we have just described that both the position and the speed of the second particle were "real" quantities, since they could both be precisely determined in indirect fashion, without disturbing the second particle. This conclusion seemed to be in conflict with the uncertainty relations associated to the position and speed of the second particle, unless there is something "magical" in quantum theory, that is to say an intimate link between systems separated by very large distances, causing every observation performed on a system to instantaneously effect the other system, and thus making it capable of changing its "real" state. Einstein, Podolsky and Rosen thought that the existence of distant links between spatially separated systems was not physically acceptable, and deduced from their reasoning that there was something incomplete in the quantum description of a system through the probability amplitude $\mathcal{A}(x_1, x_2)$ (which was the basis for their reasoning).

When it first appeared, the EPR article did not have a great impact on the community of physicists. Most of them rested their minds on the "soft pillow" of Copenhagen and took no pains to reflect on the new perspectives opened up by the EPR article. Only Niels Bohr and Erwin Schrödinger took a lively interest in this paper. Niels Bohr responded to the EPR paradox by publishing an article which essentially consisted in reaffirming the "dogma" of complementarity.[20] He thus justified what Einstein had written about him, just after the publication of the EPR article and before Bohr's response, in a letter to Schrödinger:

> As for the Talmudic philosopher, he doesn't give a hoot for "reality," that hobgoblin capable only of scaring naive souls. He explains that the two points of view differ only by their mode of expression.

Here the expression *Talmudic philosopher* refers to Bohr, thus comparing him to a commentator on the divine revelation (here understood as complementarity).

As for Schrödinger, he understood that Einstein had put his finger on an important structure in the quantum formalism. In the months following the publication of the EPR article, Einstein and Schrödinger held a discussion through the mail. In this exchange, Einstein suggested the consideration of an unstable system, like a gunpowder barrel which has a 50% chance of catching fire within a certain time. Einstein noted that after this interval of time the quantum theoretical representation of the gunpowder barrel by a probability amplitude "then describes a sort of mixture containing the system which has not yet exploded and the system which has already exploded." This suggestion by Einstein (to consider a macroscopic system whose state depends crucially on a random process) was soon taken up and improved by Schrödinger in his famous example of *Schrödinger's cat*. This is a living cat placed into a box with a diabolical mechanism which will either kill or not kill the cat within one hour, according to whether a single radioactive atom has decayed or not. At the end of an hour, quantum theory describes the cat by a probability amplitude A which corresponds to a *superposition*, with equal weight, of the amplitude for a living cat and the amplitude for a dead cat. How is this quantum description to be reconciled with the fact that we never observe such superpositions of half living and half dead cats, but only a living cat *or* a dead cat?

The heritage of the EPR argument did not stop there. In 1964, Nearly 30 years after the publication of the original article by Einstein, Podolsky, and Rosen, the Irish theoretical physicist John S. Bell took the EPR dilemma seriously, between a structure of reality called *separable*, where spatially separated systems do not influence each other at a distance, and a *nonseparable* structure where some spatially separated systems remain linked between themselves, or as is also said, are entangled, if they have had the opportunity to interact in the past. Bell understood that these two possibilities could be distinguished by certain types of measurements performed on quantum systems which have interacted in the past. More precisely, he showed that the quantum entanglement, *à la* EPR, of the "quantities of internal rotation," also called *spins* or *polarizations*, of two particles issuing from an initial state with zero spin, must lead to correlations between measurements of the polarizations of the two particles which are strictly greater in the case of a nonseparable, quantum reality than in the case of a separable, classical reality.

Bell's theoretical discovery invoked great interest in entangled situations *à la* Einstein-Podolsky-Rosen and prodded several experimental teams to test the inequalities that Bell had deduced for the correlations between the polarizations of separate particles, issued from an initially correlated system. The most convincing experimental results were realized in 1982, at the University of Orsay, France, by a group led by Alain Aspect. These results were in full agreement with the predictions of quantum theory, that is to say with a nonseparable structure of reality where two systems which have interacted in the past remain entangled in the future, even if they are spatially separated. The experiments at Orsay verified the reality of this EPR entanglement for the polarization of photons separated by a dozen meters. Some more recent experiments, completed near Geneva, Switzerland, by the group of Nicolas Gisin, have verified the reality of EPR entanglement for the polarization of two photons separated by more than ten kilometers!

Experiments performed on systems of the Einstein-Podolsky-Rosen type have thus shown that two systems which have interacted in the past continue to behave like an inseparable whole in spite of the spatial distance between them. This shows that quantum mechanical reality is very different from classical reality. In addition to leading to progress in our comprehension of quantum theory, the entangled EPR states are presently the object of numerous studies, for it is thought that they might have very important applications within the domains of quantum cryptography and quantum computation.

7

Einstein's Legacy

Einstein's life ended . . . with a demand on us for synthesis.
—Wolfgang Pauli

The Mouse and the Universe

Princeton University, Princeton, New Jersey, April 14, 1954

When the old man entered, silence fell suddenly upon the 60 or so students assembled in Room 307 of the Palmer Physical Laboratory, on that 14th of April, 1954. The students were emotional and excited. Everyone knew that it was an exceptional event, without a doubt the only time in their life that they would see, in flesh and blood, and hear the speech of the greatest physicist of all time, the living legend of twentieth-century science: Albert Einstein. They were going to attend the great man's final lecture.

Some of them had had the privilege, the preceding year, of being invited to take tea at Einstein's house, at 112 Mercer Street, and were able to hear the master's direct answers on all the questions they posed: from the nature of electricity and the foundations of the unified field theory, to the expansion of the universe and Einstein's position on quantum theory. Einstein had joined in the game with grace and good humor, and had responded in detail to all of their questions. He was not even offended when one student, bolder than the others, dared to ask him: "Professor Einstein, what will become of this house when you are no longer living?" A large smile lit up the old man's face. He replied, without becoming disconcerted, in good English with a melodious German accent: "This house will never become a place of pilgrimage where the pilgrims come to look at the bones of the saint."

The American theoretical physicist John Archibald Wheeler had begun teaching relativity (special and general) in the physics department of Princeton University starting in the fall of 1952. It had been his idea to invite the students of his course on relativity to take tea at Einstein's house, in May 1953, to help motivate them to study this theory deeply. It was he as well who convinced Einstein, in the spring of 1954, to come give a seminar before a select group of students from the physics department. Of course, the grapevine had done its work, and a fair share of students from neighboring disciplines, especially mathematics, had come to hear him. Some professors slipped in amongst the group of students which filled the small seminar room.

The central theme of this lecture—which was effectively Einstein's last seminar, given one year, nearly to the day, before his death—was quantum theory.[1] Einstein explained why he thought that this theory was not the last word on the question. He reviewed the process of an atomic transition to a state of higher energy under the influence of electromagnetic radiation. By continually lowering the intensity of radiation, this transition process becomes more and more rare. This led to the introduction of a probabilistic description of the process of transition. Thus probability was introduced into quantum theory.[2] "I am a heretic. If radiation causes jumps [between atomic states], it must have a granular character like matter," Einstein exclaimed. Then, he came to his crucial point: what is the real meaning of the probability amplitude A? Does it give a complete description of the physical situation? "I knew in constructing special relativity that it was not complete. So is everything that we do in our time: with one hand we believe; with the other, we doubt." Then Einstein gave as an example the quantum description of a macroscopic object (a sphere of one millimeter diameter moving in a box). The description of the motion, for fixed energy, of the tiny ball given by the probability amplitude seems paradoxical for an object which one can see with the naked eye. The probability amplitude gives a fuzzy description of the ball's position, while everyday experience shows that the ball is always seen at a well-defined location.

> It is difficult to believe that this description is complete. It seems to make the world quite nebulous unless somebody, like a mouse, is looking at it ... When a person such as a mouse observes the universe, does that change the state of the universe?

Many of the attendees were struck by Einstein's evocative image. Einstein then mentioned that he believed that logical simplicity could, sometimes, be a good guide, for it is thus that he had constructed the theory of general relativity. He explained how he had found this theory, and why he thought it was incomplete: the description of matter by means of the distribution of energy and stress seemed to him to be something provisional, "a wooden nose in a snow man." He regretted that most physicists took quantum theory and the theory of special relativity as their starting point, while neglecting gravity as being unimportant. On the contrary, he thought that gravitation, or more generally the structure of space-time, must be taken into consideration from the beginning. He finished by indicating that: "There is much reason to be attracted to a theory with no space, no time. But nobody has any idea how to build it up."

Among Einstein's audience, on April 14, 1954, was an emaciated, nervous young man with an eagle's profile and an intense gaze: Hugh Everett III.[3] He was only 23 years old, and had come with his friend Charles Misner, who was taking Wheeler's course in relativity. Hugh Everett would not have missed this opportunity to hear his idol for anything in the world. At 12 years old, he had written to Einstein to ask him whether the universe was based on a structure that was random or unified. And he had had the great surprise of receiving a friendly reply from Einstein himself. After having studied chemical engineering for the first two years of university in Washington, DC, he had spent the last six months (since September 1953) at Princeton University, where he was affiliated with the mathematics department. However, he was, in fact, interested primarily in theoretical physics. Since classes had commenced in September 1953, he had followed, in particular, the course on introductory quantum mechanics given by Robert Dicke.

Hugh Everett was struck by Einstein's remarks on the apparently incomplete character of quantum theory, which offered a nebulous description of the universe, and which seemed to need the presence of living beings, even if it only be one mouse, to trigger what the partisans of the Copenhagen dogma called the collapse of the wave-packet: the passage from a fuzzy world to the sharply defined world that we see around us. He began to seriously reflect on the physical meaning of the formalism of quantum theory.

Some months later, during an evening party soaked with sherry, an animated discussion took place at the Graduate College between Hugh Everett, Charlie Misner, and Aage Peterson, who was an assistant of Niels Bohr, and who was passionately interested in the problems posed by the interpretation of quantum theory. In the heat of conversation, Hugh sketched a new conceptual scheme for the interpretation of quantum theory in such a way as to avoid both the paradoxes raised by Einstein (and Schrödinger), and the necessity of assuming a mysterious random process of wave-packet collapse. This idea of genius, obtained when he was about 24 years old, was the seed of Hugh Everett's doctoral thesis, in which he developed a revolutionary interpretation of quantum theory.

Everett went to see John Wheeler (who had been a disciple and collaborator of Niels Bohr, and who was very interested in the meaning of quantum theory) and asked him to supervise his doctoral thesis. Wheeler accepted. This created some problems for Everett. On the one hand, Wheeler was quite open to new ideas, and he encouraged his students to think for themselves. On the other hand, Wheeler had unconditional admiration for Bohr and his principle of complementarity. Because of this, while recognizing the innovative character of Everett's ideas, Wheeler presented many objections to the way in which they were expressed. For example, in a note to Everett from September 1955, Wheeler wrote that he would be "frankly bashful about showing it to Bohr in its present form" since it could be "subject to mystical misinterpretations by too many unskilled readers." Finally, after insistent advice from Wheeler, Everett summarized the long text in which he developed his ideas in detail into a much shorter text which he defended as a doctoral thesis in 1957, and which was published the same year, accompanied with an assessment by Wheeler.

Everett's interpretation of quantum theory is one of the great conceptual advances of twentieth-century physics. The author of this book thinks that it would have pleased Einstein (who died when Everett had just begun to develop his idea). Indeed, not only did it supply a new response to the paradox of the mouse looking at the universe, mentioned by Einstein in his final lecture, but above all it fits perfectly with Einstein's scientific philosophy, such as we have previously outlined it. Let us recall Einstein's statement to Heisenberg, "The theory itself defines what is observable," which put Heisenberg on the path to one of the first conceptual advances in quantum theory: the uncertainty relations. As we shall see, Everett's interpretation is the first to take Einstein's statement seriously.[4]

Nevertheless, in spite of—or, perhaps, because of—its novelty, Everett's interpretation raised no interest. Before it was revived, through the efforts of the theoretical physicist Bryce DeWitt in the 1970s, it was completely ignored, even by the recognized experts on the history of quantum mechanics (like Max Jammer). This rejection is doubtless due in part to the total lack of interest in Everett's ideas shown by Niels Bohr himself. Bohr read the long version of Everett's thesis, and raised some objections. In the spring of 1959, at Wheeler's insistence, Everett visited Copenhagen for six weeks in order to meet Bohr and discuss his interpretation with him. Everett had a very bad memory of this meeting. Bohr was not interested, and he gave Everett no opportunity to explain his ideas in detail.[5] Today, according to a recent poll conducted by email, the majority of theoretical physicists interested in understanding cosmology within a quantum framework use Everett's interpretation. In fact, they have no choice. As written recently by Bryce DeWitt, who rescued Everett's interpretation from obscurity:

> Everett's interpretation has been adopted by the author [Bryce DeWitt] out of practical necessity: he knows of no other. At least he knows of no other that imposes no artificial limitations or fuzzy metaphysics while remaining able to serve the varied needs of quantum cosmology, mesoscopic quantum physics, and the looming discipline of quantum computation.[6]

The Multiple World

What is the essential idea of Everett's interpretation? To introduce it, let us recall the central paradox of quantum theory, such as was highlighted by the arguments of Einstein's gunpowder barrel (half-exploded, half-intact) and Schrödinger's cat (half-living, half-dead). Quantum theory describes the system consisting of the cat and its environment (the box enclosing it, the air it breathes, the lethal mechanism triggered by a radioactive atom, etc.) by a function of the configuration of the system. To each configuration q of the system is associated a (complex) number $\mathcal{A}(q)$ that we shall simply call the amplitude of configuration q. What is a configuration q, considered at a fixed time t, and how is it described? For example, one could describe each possible instantaneous configuration of the cat and its environment by specifying the position in space of each of the system's atoms[7] (the atoms making up the cat, those in the air, those in the lethal mechanism, etc.). The position of each atom is specified by giving its three coordinates in space (length, width, and height). Let N be the number of atoms in the system. The number N is gigantic. Indeed, we recall that a gram of matter contains around 600 thousand billion billion (6×10^{23}) atoms. A configuration of the total system is thus specified by giving a (gigantic) list of $3N$ numbers. The notation q denotes such a list.[8]

Dear reader, maybe you take fright at the thought of considering a quantity $\mathcal{A}$ which depends on such a gigantic number of variables. All the more so since, as we have briefly indicated, the amplitude $\mathcal{A}$ is not a regular "real" number (like 2.5 or 3.1416) but a complex number, which is to say essentially an arrow within a plane, which requires two real numbers for its description (for example, the length of the arrow, and the angle that it makes with an arrow pointing east). To visualize what such an amplitude $\mathcal{A}$ means, we can use a representation introduced by the author in a previous book.[9] It consists of (mentally) using the techniques of film-making.

First, each configuration q of the system is represented by a (holographic[10]) photographic image of the system at the instant considered. To each q, that is, to each photographic image of the system, we want to associate a certain amplitude $\mathcal{A}$, determined by an arrow in a plane, having a certain length and pointing in a certain direction. To each direction of the arrow, one may associate a particular hue of color on the "color wheel": for example, we associate to the direction east (on a map) the color orange, then, as one rotates the direction in a clockwise direction, one changes the corresponding color by passing successively from orange (east) to red (southeast) to violet (south) to indigo (southwest) to blue (west) to blue-green (northwest) to green (north), and finally to yellow (northeast). As we continue to rotate from northeast to east, the hue evolves continually from yellow back to orange, in such a way that we land again solidly on our feet, having spread a full spectrum of hues around the circle. We have already mentioned that each amplitude $\mathcal{A}$ corresponds to a length and a direction. To the length, we can associate an intensity of light (a weak intensity if the length is short and a strong one if the length is long), and to the direction a hue of color (for example, orange). Thus we can fix each complex amplitude $\mathcal{A}$ by a color, having both a particular intensity and a particular hue: for example, a high-intensity orange, or a medium-intensity red, or a weak-intensity green, etc.

Let us then combine these two representations: that of the spatial configuration of the system by a photographic image (initially in black and white), and that of the amplitude $\mathcal{A}$ associated to this configuration by a color (intensity and hue). This gives us a photographic image having a certain intensity and a certain hue. For example, at a given instant, the living cat with his environment is represented by an intense blue image, and the dead cat with his environment by a red image of the same intensity. We may now superpose these two images, by the film-making technique of double exposure (see Figure 13). That is to say, we print onto the same frame the two preceding images. This multiple exposure of images of the system, colored more or less intensely, gives a fairly faithful representation of the mathematical notion of a complex amplitude $\mathcal{A}$ depending on a spatial configuration q. To complete this representation, it suffices to vary the instant t at which we consider the system. Thus, to each instant t there corresponds a frame, multiply ex-

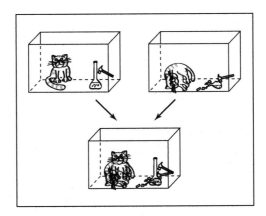

Figure 13. Schrödinger's cat and quantum "double exposure."

posed to several colored images with more or less intensity. By considering all the successive instants, we thus obtain a (continuous) series of (colored and multiply exposed) images, that is to say a film, in color and with multiple exposures. Finally, we must imagine that the hue of each configuration changes extremely quickly, moving rapidly around the color wheel, as soon as the configuration is modified, even in an infinitesimal fashion (for example, as soon as a single atom of the configuration moves). Moreover, even for an empty frame, where the configuration does not move at all, we must imagine that its hue changes very rapidly in the course of time, rotating at top speed around the color wheel (while the intensity of the light remains constant).[11]

Let us now explain Everett's idea. It consists in taking seriously Einstein's statement: "The theory itself defines what is observable." Let us first take quantum theory seriously and ask it to define what is real. Each configuration q will have more or less "reality" according to the value of the amplitude $\mathcal{A}(q)$. In other words, we interpret $\mathcal{A}$ as an *existence amplitude*, and not (like in the Born-Heisenberg-Bohr interpretation) as a *probability amplitude*. Indeed, the notion of probability amplitude for a certain configuration q suggests, from the very beginning, a random process by which only one configuration, among an ensemble of possible configurations, is realized, passing from the possible to the actual. By contrast, the notion of existence amplitude suggests the simultaneous existence (within a multiply exposed frame) of all possible configurations, each actually "existing," but with more or less intensity (with the color encoding the "orientation" of the amplitude $\mathcal{A}$, which in physics is called its phase).

Using the film-making analogy explained above, let us now describe the two basic elements making up the Everett interpretation. The first consists in saying that quantum reality is a color film with multiple exposures. At each instant, all of the individual images in the exposure "exist" with an intensity given by the length of the complex amplitude $\mathcal{A}$. The only configurations q which do not exist are those with a null amplitude $\mathcal{A}(q) = 0$. Having arrived at this stage, the reader may say to himself that the film obtained by successively projecting all these multiply exposed images will be effectively invisible. It will only offer an infinite jumble of confused images. We seem to thus recover the nebulous or fuzzy description of which Einstein and Schrödinger complained, while we actually see, around us, reality "existing" in one well-defined configuration, like in a unique film with sharp images and no double exposures.

It is here that the second element of Everett's interpretation comes into play. To completely explain this second element, we must first have recourse to certain mathematical characterizations measuring the fact that certain images (or certain successions of images, that is to say certain films) are so different from each other that, when we superimpose them, they "don't interfere" with each other, with the effect that we can focus on one image or the other. We allude here to a mathematical phenomenon similar to what is known as the cocktail party effect: the possibility for two people to have a conversation between themselves, in the middle of the brouhaha formed by the intersecting conversations of other people. Another analogy, helpful for radio owners, would be that of changing the reception frequency to be able to listen, without interference from the other channels on the dial, to one particular station.

In other words, to return to our cinematic analogy, Everett tells us that among the hodge-podge of the total multiply exposed film, there exist subfilms with (more or less) sharp images, which evolve in time according to (more or less) logical scenarios. The important point here is that the characters who evolve within such a subfilm act, at each instant, under the influence (almost) exclusively of those things which they have seen or felt in the previous images of the same subfilm.

Let us give a cinematic example of this idea. In the middle of Frank Capra's beautiful film *It's a Wonderful Life*, the hero, George Bailey, played by Jimmy Stewart, wants to commit suicide on Christmas Eve, because he believes himself to be a useless failure. Clarence the angel then plays out before his (and our) eyes, from the beginning, the film of what would have happened if George had never existed. This second film also develops in a coherent manner, and progressively becomes quite different from the first, which is to say the first half of Capra's film. Everett's idea is essentially that, in the total quantum reality, the two halves of the film (with or without George Bailey) are superimposed on each other, and thus play out simultaneously. Nevertheless, within each subfilm each character only has knowledge of what has happened and is happening in their own layer of the film, and thus has no consciousness of the existence of the other subfilm, playing out on a neighboring layer.

Let us finally note that Everett did not completely establish the necessity of what he proposed. By making the hypothesis of the existence of subfilms, which do not interfere with each other, he realized an essential desideratum of Einstein (*Probabilitatem esse deducendam*), that of justifying the connection between the existence amplitude $A(q)$ and the probability for an observer to see the corresponding configuration q.[12] Later, other physicists justified the (apparent) existence of subfilms which do not interfere with each other by studying what is now called the *decoherence* between two possible subfilms.[13]

We further note that Everett's interpretation was called, by he who brought it back from obscurity, Bryce DeWitt, the Many-Worlds Interpretation. This name refers to the existence of numerous noninterfering subfilms in the midst of the total, multiply exposed film. One then says that the world splits at every moment into multiple, slightly different versions which, in their turn, split in the following instant, etc. This leads to an image of a world which continually branches out in a multiplicity of separate worlds. This image has been used by excellent physicists who well understood Everett's interpretation: notably Bryce DeWitt and David Deutsch.[14] I nevertheless find this image inappropriate, as it suggests a complete splitting between separate classical worlds, similar to the splitting from one cell into two, and so on to an irreversible multiplication. I prefer to remain closer to the formalism of the theory itself and to speak of a multiple world, that is, one film, multiply exposed.

We mention finally that by calling reality a *multiple world* one could (and in fact one should) understand the word *world* in the sense used by Minkowski, which is to say a space-time. Classical (in the sense of prequantum) relativistic reality is identified with a unique space-time, that is to say with a four-dimensional world. In our cinematic analogy, such a world corresponds to one film: a sequence (or a "stack") of three-dimensional images. Quantum reality corresponds to a multiply exposed film, which is to say a stack of superimposed images. Note that, starting from this stack, one can

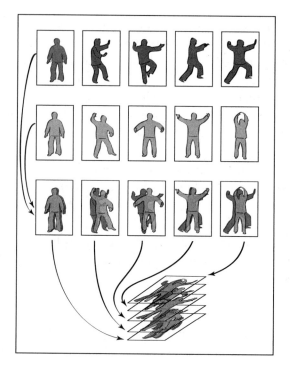

Figure 14. The quantum world as a multiply exposed film.

distinguish *a priori* a very large number of subfilms, many more than the number of layers of exposure within one instantaneous image. Indeed, if one considers a mini-film of three successive images, each of which has two layers of exposure, one can assemble $2 \times 2 \times 2 = 2^3$ subfilms of three successive images, each of which is taken at random from the two possible images at each of the three instants of the total film. Everett nevertheless tells us that most of these subfilms only exist with an amplitude too weak to be perceived. Only certain quasi-classical subfilms, whose amplitudes are reinforced by a process of constructive interference, will exist with an amplitude strong enough to be perceived.

The Kantian Quantum

The reader may be saying to himself that Everett, and those who adopt his point of view, have truly passed outside the boundaries of the reasonable, and that this idea of a multiple, phantasmagorical world is too absurd to be taken seriously. It is indeed because of the revolutionarily "absurd" character of Everett's idea that it was ignored (particularly by Bohr), rejected, or considered taboo for nearly 30 years. Even today, some experts in the interpretation of quantum mechanics reject, with an incredulous and disapproving sneer, Everett's interpretation by arguing that it shamefully violates the principle of logical economy proposed by William of Ockham: "One should not increase, without necessity, the number of entities required to explain something."

On the contrary, we would like to point out that Everett's interpretation is characterized by its logical economy. It is the only interpretation of quantum theory which does not add foreign (physical or metaphysical) elements to the theory. As for us, we consider that it is the only possible interpretation (see also the above quote by Bryce DeWitt) and that it finds its justification in the most rigorous and most rational epistemology, notably that of the German philosopher Immanuel Kant.

One of Kant's goals was to clarify the nature of the objects (space, time, force, and matter) which science speaks of, and to understand to what extent science makes true assertions about these objects. For example, is the absolute space evoked by Newton something "real," which exists by itself independently of things? Is the Euclidean geometry that is attributed to space true in an *a priori* fashion, before making measurements to verify it? This is not the place to discuss in any detail the responses Kant gave to these questions.[15] Let us simply say that while Kant of course recognizes the essential role of experimentation in the progress of physics, he strongly insists on the fact that experimentation is only truly fruitful if reason takes the forefront by posing a logical and mathematical framework permitting the interpretation of experimental results, and giving them meaning. This concept upends the very notion of reality, that is to say the meaning of what is an object or a thing for the rational investigator. As Kant wrote:

> Hitherto it has been assumed that all our knowledge must conform to objects. [...] Let us try to see whether we may not have more success in the tasks of metaphysics, by assuming that objects must conform to our knowledge.

Let's apply this philosophy to the interpretation of quantum theory. This will lead us to what I like to call the Kantian Quantum,[16] where the word *Kantian* makes reference to a perfectly rational approach. The Kantian Quantum is thus an approach in which we must adjust our understanding of "objects," that is to say the very notion of reality (the word *reality* derives from the Latin *res* = thing), to our knowledge, or, more specifically, to the quantum theory itself.

Indeed, quantum theory has been verified by a huge number of experiments which have, in particular, confirmed the validity of its most "bizarre" consequences, like the entanglement predicted by Einstein-Podolsky-Rosen between separated systems, and the superposition of different macroscopic states, of the type of Einstein's gunpowder barrel or Schrödinger's cat.[17] One can, and in fact should, until getting contrary information, consider quantum theory as firmly established knowledge. Then, starting from this knowledge, that is, from the mathematical formalism of quantum theory, if we ask this formalism to define the nature of quantum objects, or quantum reality, one necessarily falls back on Everett's point of view, since it is the only "interpretation" which is founded uniquely on the formalism of the theory, adding neither "fuzzy metaphysics," nor verbal incantations, nor new, nonverified hypotheses.

I have stated several times that Einstein himself had asserted his adhesion to a point of view close to Kant's ("Only the theory decides what is observable"). It is of interest to note that he expressed himself in a letter to Schrödinger (written just after the appearance of the EPR article) in a manner very close to Kantian views, and precisely in regard to the mysterious character of quantum reality:

> The true difficulty lies in the fact that physics is a sort of metaphysics; physics describes "reality." However, we do not know what "reality" is, we only know it through the description given by physics!

The Grand Illusions

If one reflects on the upheavals brought to the notion of reality by the physics of the twentieth century, through the essential driving force of Einstein, the account is almost unbelievable. At the end of the nineteenth century, reality (such as it was defined by the science of the time) seemed quite familiar, and close to the common concepts that every individual had about the world around him.

There was Space, the great scene where the theater of reality played out. This containing space had nothing to do with the material contents of the universe. Its geometric properties were those that the Greeks had discovered two millennia before, and that everybody learned in school.

There was Time, the vast cosmic pulse present in all the universe, which seemed to coincide with the duration experienced by all human beings. Was it not existentially evident that time passes?

There was Matter, which was defined by its permanence. It was the eternal substance of the world: "Nothing is destroyed, nothing is created."

There was Force, which caused the evolution, with the passage of time, of matter within space.

And of course, all of this gigantic clockwork mechanism only existed, by definition, in one single example. The Universe, the ensemble of all reality present in infinite space, was unique. How could it be otherwise? Do we not see, at every instant, a single world surrounding us?

All of this "self-evidence" was mercilessly crushed under the knock-out punches of the ideas initiated by Einstein. The passage of time is an illusion. The reality defined by special relativity is a four-dimensional space-time block where nothing corresponds to any passing of time whatsoever, and where it is impossible to define a "now." Special relativity also tells us that matter is ephemeral and can be created from energy, or destroyed into impalpable radiation. General relativity then tells us that the space-time block is, in fact, a space-time-force-matter block where the contained force-matter is entangled in a quasi-indissociable way with the containing space-time, which it deforms by its presence. Finally, quantum theory tells us that this space-time-force-matter block is not unique, but multiple. And this multiplicity is not that of a juxtaposition of separate spatio-temporal worlds, but that of a superposition, within the same reality, of an infinity of spatio-temporal worlds which co-exist, in the manner of a multiply exposed film.

We leave it to other books to explain how one may reconcile the everyday evidence, concerning in particular the passing of time and the uniqueness of reality, with the new reality defined by the theories of relativity and quantum theory.[18] Let us here simply state that one of the crucial ingredients to effectuate such a reconciliation is to account for the nature of the physical signals which underlie the consciousness we have of being in the world. For the illusion of time, the irreversible nature of memory encoding plays an essential role. For the illusion of the uniqueness of reality, the relatively crude (coarse-grained) nature of the information which our senses give us about our surroundings plays the essential role, along with the phenomenon of noninterference between configurations which are just a tiny bit different, macroscopically.

Dreams of Unification

We have already evoked the program of unification which Einstein pursued for about 30 years. This program failed, in the sense that he never obtained a concrete result capable of transcending general relativity and quantum theory, and explaining the necessity for the various interactions of physics. Nevertheless, it is remarkable that a theory presently in development seems to accomplish, through completely new methods, some of Einstein's dreams of unification. This theory is called, in a provisional fashion, string theory, since its point of departure consists in saying that the elementary building blocks constituting matter (and force) are strings, and not, as had been previously supposed, point-like particles.

Recall that for 2,000 years, ever since the Greek atomists had divided everything into atoms and the void, matter was conceived as being constituted of point-like corpuscles. Western physics added the notion of force, and showed that it derived from a continuous field created by matter. One may think, for example, of the electrical field created by charged particles. But in 1905 Einstein introduced the revolutionary idea that the electromagnetic field was, in fact, made up of point-like corpuscles, the quanta of light, or photons. The work of Heisenberg, Jordan, Dirac, Pauli, and others in the 1930s showed how the mathematical formalism of quantum theory permitted the reconciliation of the apparently contradictory aspects of continuous field and discontinuous particle. Then it was shown, thanks in particular to the work of the American physicist Richard Feynman, that one could reconstruct the entire theory of quantum fields by applying the basic postulates of quantum theory to the dynamics of point-like particles.

The theory of quantum strings is then defined by applying the same basic quantum postulates to the dynamics of relativistic elastic strings. Such a string is analogous to a rubber band, or a thin rubber ribbon. A string may be closed, that is to say closed in on itself in a loop, or open, with two separate ends. A relativistic string possesses an internal tension which tends to shrink its length. More precisely, this tension, if it acted alone, would reduce the length of the string to zero. Contrary to an ordinary rubber band which possesses, when left alone, a nonzero length, and which only develops tension if one stretches it, a relativistic string is always under tension, and its "length at rest" is zero. Because of this, a relativistic string can only acquire nonzero length if it is not at rest, but on the contrary is agitated by an incessant motion. For example, it can regularly turn on itself like an ice skater who turns with arms extended, or else it may move by shaking in every direction like a break-dancer.

We shall not enter here into the details of the quantum theory of relativistic strings. This theory was initiated in 1968 in work by Gabriele Veneziano. It was subsequently developed over the next 30 years through the work of numerous physicists, notably Veneziano, Miguel Virasoro, Pierre Ramond, André Neveu, John Schwarz, Joël Scherk, Michael Green, Alexander Polyakov, and David Gross. It was then understood, notably by Paul Townsend, Joseph Polchinski, and Edward Witten, that string theory also contained more complicated extended objects. These objects are *elastic membranes* (like a rubber balloon) or, more generally *p-branes*: objects extended in a number *p* of spatial directions.[19]

The two fundamental pillars upon which string theory is constructed are the (1905) special theory of relativity and the quantum theory. The original formulation of string theory completely ignores the general theory of relativity. Nevertheless, quite remarkably, it can be shown that string theory contains, as a subsector, the theory of general relativity. This is quite surprising, for string theory presupposes, as its point of departure, a sharp separation between a rigid container (Minkowski space-time) and the elastic contents (the strings). But the arrival point of the theory is that the contents have, in a sense, transferred part of their elasticity to the container, which becomes the elastic space-time of general relativity.

From this point of view, string theory (partially)[20] realizes one of Einstein's ideas, that in which gravitation, described as a space-time deformation, is not an accessory element of reality, but is rather something indispensable, which must play a fundamental role. Moreover, one finds that string theory predicts that the geometric structure of space-time is richer than the one used in general relativity. In a surprising way, it is found that some of the new geometric structures suggested by string theory are connected to Einstein's "last unified theory,"[21] on which he worked until his final breath.

Another of Einstein's ideas was to unify (Maxwell's) electromagnetic field with the gravitational field (in Einstein's sense). It has often been said that Einstein's hope was useless and naive. Nevertheless, in a surprising way, string theory seems to deeply unify electromagnetic interactions (and their generalizations, called gauge or Yang-Mills interactions) with Einstein's gravity. This unification is still mysterious, but it is thought to contain an important key for the future of the theory.[22] We shall here only note that, in broad outline, the electromagnetic field is connected to open strings (with two ends), while the gravitational field is connected to closed strings (that is, closed into a loop).

Einstein also hoped to eliminate the point-like singularities which appear in Minkowski space-time when one considered the fields created by point-like sources. He thought that gravitation could replace these singularities by regular zones, like the Einstein-Rosen bridges that he had studied in 1935. Once again, string theory seems to realize this hope in a roundabout fashion. Indeed, some recent work[23] in string theory has exhibited a deep and mysterious equivalence between sources of certain fields, analogous to the electromagnetic field, and a deformed space-time. When gravitation is neglected, these sources (called Dirichlet branes) engender singularities of the field. However, when the effects of gravitation are important, the deformed space-time which they produce is everywhere regular. In addition, these deformed space-times contain geometric structures of the same type as Einstein-Rosen bridges. Finally, in a truly remarkable way, the equivalence we have just spoken of permits the identification of certain processes or results of a typically quantum nature, to nonquantum, geometric phenomena.

As we see, many of Einstein's hopes thus find an unexpected realization within work on the very frontier of theoretical physics. We must nevertheless take note of the fact that the context in which these hopes are in part realized is very different from that which Einstein had envisioned. In particular, it is essential to take quantum theory as the point of departure in order to arrive, in string theory, at the phenomena we have just described.

Not a Day without Einstein

Einstein once lightheartedly said: "They should all be ashamed, those who thoughtlessly make use of the miracles of science and technology, without understanding any more of them than the cow does of the botany of the plants it eats with enjoyment." Elsewhere he insisted on the fact that the original source of all technical accomplishments is "the divine curiosity and the playful urge of the tinkering and pondering researcher." Out of respect for Einstein, I would thus like, dear reader, for you to think sometimes of all the commodities and technology of everyday life which have been initiated through Einstein's playful urge to ponder on the structure of reality.

In fact, we note that Einstein was not a pure theorist, having no direct interest in practical applications. Throughout his life, beginning with his laboratory studies in Zurich and his work at the Patent Office, he kept an interest in experimental research and scientific applications. (Recall as well that his uncle Jakob was an engineer who worked, with his father, to develop the electrification of the city of Munich, and then of the Pavia region of Italy.) In particular, Einstein himself applied for a fair number of patents for various inventions, ranging from an apparatus for measuring small voltages, to hearing aids, a silent refrigerator, and magnetically levitated gyro-compasses.

Physics is omnipresent in our environment, and a great part of modern physics has directly issued from the theoretical ideas formulated by Einstein. Lasers have multiple practical applications: from industrial cutting to compact disc players, as well as the manipulation of individual biological molecules and a plethora of various guidance systems. Let us then reflect from time to time that it is one of Einstein's insights from 1916, on the exchange of energy and momentum between atoms and quanta of light, which led to the prediction of the process which is the very base of the laser: the stimulated emission of radiation.

If you have not listened to a compact disc today, and have thus missed the opportunity to think of Einstein's 1916 work, will you perhaps watch television? Think then that the electrons accelerating within the cathode ray tube acquire a speed on the order of one-third of the speed of light, and thus that the precise control of their path towards the screen necessitates the consideration of the dynamical equations of the theory of special relativity, obtained by Einstein in June 1905.

But perhaps you have decided not to stay at home, and instead will go run some errands. It is then quite probable that you will pass through doors which open through the use of photoelectric cells. Think then of the fact that the fundamental theoretical law of the photoelectric effect was obtained by Einstein in March 1905. Think also that this law was not discovered in view of its applications, but as a byproduct of a deep reflection on the nature of light.[24]

Perhaps you will take a taxi guided by the Global Positioning System (GPS)? Think then that Einstein's general theory of relativity is used in a crucial fashion in this system, which today has an ever-growing number of applications, from guiding airplanes and ships to guiding tractors through immense fields with centimetric precision. Indeed, the GPS system is based on the transmission, to the user, of time signals emitted by atomic clocks in orbit around the earth. The software of the GPS system takes the space-time deformation caused by the mass of the Earth into account. This deformation causes the clocks traveling in the satellites to seem, as seen from the ground, to run more quickly (to which is also added the effect of the orbital speed which, according to the theory of special relativity, implies that the traveling clocks seem, as seen from the ground, to run less quickly). The two effects do not cancel each other out, and are both, although minuscule, quite important with regard to the precision of chronometry required for the system to work.[25] If one did not account for the predicted effects of the two theories of relativity, the GPS system would be unusable after a few minutes.

One could look at other examples as well. If one remembers that the first physicist to take quanta seriously—not only for light, but also, as we have seen, for the possible energy levels of material oscillators—was Einstein, one could think about him in relation to the many applications of quantum physics, notably in solid-state physics.

Finally, there is an interesting study of those scientific articles published before 1912 which were the most cited between 1961 and 1975 which showed that among the 11 most influential articles, four were due to Einstein alone, and the seven others had been written by seven different authors.[26] What is even more remarkable, is that this list of the most influential scientific articles of the century contain neither the article of March 1905 on the quanta of light, nor the article of June 1905 on relativity. In fact, these two articles have been so influential and important, they have so deeply modified the course of twentieth-century science, that no one cited them any more 50 years after their appearance! They had become the very flesh of modern physics. Because of this, one discovers with surprise that others of Einstein's articles, and notably the other articles which he wrote during the "miraculous year" of 1905, but of which we have not spoken above, are found in the list of articles having had the greatest impact. Indeed, in 1905, Einstein wrote his thesis concerning a new determination of molecular dimensions and an innovative article on the "Brownian motion" of pollen grains, or colloidal particles, suspended in a liquid. These works have found, and still find, application within a vast domain of research ranging from ecological work on the dispersion of aerosols in the atmosphere to studies on the behavior of the casein particles in milk during the process of cheese fabrication!

Of course, outside of the incredibly vast panoply of technological applications of Einstein's work (from the laser to the fabrication of cheese), one must keep in mind the complete upheaval of the conceptual landscape of science brought by his theories: from the Big Bang to the quantum states of a photon, passing through black holes, binary pulsars, gravitational waves, dark energy, the manipulation of individual atoms by radiative transitions, (Bose-)Einstein condensates, Einstein-Podolsky-Rosen entangled states, and so forth. The sciences and technologies of the early twenty-first century no longer have much in common with those of the beginning of the twentieth century, which were founded on Newtonian mechanics and the thermodynamics of Carnot and Clausius. The technological revolutions of the twentieth century have had as their source the new physical theories created or initiated by Einstein and others. We are probably on the brink of new technological leaps, for example, those connected to new advances in quantum physics. Let us keep in mind that every technical accomplishment flows, as Einstein said, from the "divine curiosity and the playful urge of the tinkering and pondering researcher."

Of Bears and Men

To conclude, we pass the floor to Einstein. The essential aim of this book is to help the reader partake in the intellectual joy that scientists feel in exploring the conceptual universe opened up by Einstein and the other initiators of present-day physics. Its message is addressed in particular to young readers, whose high school courses may not always reveal the fundamental beauty of the basic concepts of science. Let's listen to what Einstein said, shortly after his arrival in the United States in 1933, to the first-year students of Princeton University:[27]

> If an old student may say a few words to you they would be these: Never regard your study as a duty, but as the enviable opportunity to learn to know the liberating influence of beauty in the realm of the spirit for your own personal joy and to the profit of the community to which your later work belongs.

To this, we may add some practical advice, for the youth who ask themselves how to contribute to the augmentation of "the beauty of the realm of the spirit." At the beginning of his career, while still employed in the Bern Patent Office, but having already made major contributions to science, Einstein would often receive visits from scientific colleagues. These people asked him how, in the middle of the setting in which he worked, he was able to have such revolutionary ideas, whose innovative character had surprised all of his colleagues. Einstein would take them to visit the very symbol of the town of Bern, the Bear's Den.[28] At feeding time, Einstein pointed out to his visitors how the great majority of bears remained pressed together, on four legs, snouts to the ground, always pacing in the same corner of their den while searching for any food fallen at their feet. Meanwhile, from time to time, a slightly different bear would rear up on its hind legs, and would look to the distance, to search for morsels of food which might have fallen farther away or in a less accessible place.

Notes

Chapter 1: The Question of Time

[1] What follows is a hypothetical reconstruction of the day during which, in conversation with his friend Michele Besso, Einstein conceived the key idea that would allow him to create the theory of relativity. Einstein's writings only indicate that this happened on a beautiful day in mid-May, five or six weeks before June 30, the date on which his article was received by the *Annalen der Physik*. If this occurred on a Sunday (Einstein's only free day; he worked eight hours a day, six days a week at the patent office), it would probably have been the 21st of May 1905 (without excluding the possibilities of the 14th or 28th of May). We note that, according to Albrecht Fölsing's excellent biography of Einstein (see the bibliography), Einstein had moved on Monday, May 15 into a new apartment, close to Besso's and in the outskirts of Bern.

[2] Einstein's work on "Brownian motion" rapidly received a favorable welcome in the scientific community, and would be, for a long while to come, his most cited work.

[3] Since 1676. In that year, the Danish astronomer Ole Roemer, working in Paris with observational data obtained by the Italian Jean-Dominique Cassini (whom Louis XIV had brought to France to direct the Observatoire de Paris), had shown that light propagated at a finite speed, on the order of 300,000 km/s. The exact value of the speed of light is (by definition) 299,792,458 meters per second. To simplify, we shall round this to 300,000 kilometers per second, equal to about 186,000 miles per second.

[4] In 1864, Maxwell summarized his research on the combined evolution of electric and magnetic fields in one system of equations that intimately mixed the two fields. After Maxwell's work, one spoke of the theory of "electromagnetism," referring to the coupled dynamics of the two fields.

[5] In the general case, this *sum* must be understood as the sum of two vectors.

[6] *The sun shines for everyone* (subtext: for eagles as well as sparrows)

[7] English translations of Einstein's most important papers on relativity, along with some important work by other authors, can be found in *The Principle of Relativity*, with notes by A. Sommerfeld, New York, Dover (1952). See also J. Stachel's recent book, *Einstein's Miraculous Year*, Princeton, Princeton University Press, 1998, which contains English translations of Einstein's revolutionary papers of 1905, accompanied with introductions and commentary.

[8] The evolution of the concept of space is described in Max Jammer, *Concepts of Space*, New York, Dover, 1993, and in Alexandre Koyré, *From the Closed World to the Infinite Universe*, Baltimore, Johns Hopkins Press, 1957. The evolution of the concept of time is discussed in Paul Davies, *About Time*, New York, Touchstone, 1996, and in Étienne Klein, *Chronos: How Time Shapes Our Universe*, translated by Glenn Burney, New York, Thunder's Mouth Press, 2005.

[9] Isaac Newton, *Philosophiae Naturalis Principia Mathematica*, London, 1687 (publishing approval dated July 5, 1686). Here we follow the English translation (from the original Latin) of Andrew Motte, London, 1729.

[10] H. G. Alexander, editor, *The Leibniz-Clarke Correspondence*, Manchester University Press, and Barnes & Nobles, New York, 1984. In their correspondence, Samuel Clarke essentially played the role of Newton's substitute. Although Clarke wrote in English, Leibniz, as a German writing to an Englishman, wrote in the courtly French used throughout Continental Europe.

[11] Written at the request of Paul Arthur Schilpp, editor of a book published in 1949 for Einstein's seventieth birthday. The English translation (from the original German) is also published separately as *Albert Einstein: Autobiographical Notes*, ed. Paul Arthur Schilpp, Peru, Illinois, Open Court, 1979.

[12] Einstein and two of his friends, Maurice Solovine and Konrad Habicht, would meet regularly in the evenings, around a humble meal, to read and discuss philosophical or epistemological works. They lightheartedly referred to this informal discussion club as the Olympia Academy.

[13] That is, these waves were both vectorial and transverse.

[14] More precisely, visible light is composed of electromagnetic waves with wavelengths between 0.4 and 0.8 microns, one micron being one-millionth of a meter, or one-thousandth of a millimeter.

[15] As has already been noted, the experimental and theoretical work of Hertz in 1887 established in a definitive way the identification between light and electromagnetic waves. Maxwell, who died in 1879 at age 48, did not live long enough to see the triumph of one of his most remarkable discoveries.

[16] In fact, the ideas held by physicists of the late nineteenth century were more varied than that. To simplify our discussion of Einstein's contribution, we shall suppose that everyone identified the ether with Newton's absolute space, at rest.

[17] It is interesting to note that in a letter to his friend Konrad Habicht, to whom we alluded at the beginning of the chapter, the only paper he identified as revolutionary was the March 1905 article on light quanta. With regard to his June 1905 article on relativity, he contented himself with remarking, "It is about an electrodynamics of moving bodies that is built upon a modification of the theory of space and time. The purely kinematical part of this work will surely interest you."

[18] Notably Henri Poincaré and Emil Cohn. See the detailed study by science historian Olivier Darrigol: "The electrodynamic origins of relativity theory," *Historical Studies in the Physical Sciences*, **26**, 2 (1996).

[19] See, for example, E. T. Whittaker, *A History of Aether and Electricity*, London, Nelson, 1953. For more recent examples of such work, in French, see Jean-Paul Auffray, *Einstein et Poincaré*, Paris, Éditions Le Pommier, 1999; Jules Leveugle, *La Relativité, Poincaré et Einstein, Planck, Hilbert*, Paris, L'Harmattan, 2004; Jean Hladik, *Comment le jeune et ambitieux Einstein s'est approprié la relativité de Poincaré*, Paris, Ellipses, 2004.

[20] Independently of the speed of the source.

[21] All of the clocks used by Einstein in the various reference frames under consideration are assumed to be "of identical construction," such that if one places them side by side, motionless with respect to each other, they tick at the same frequency.

[22] The use of telegraphic signals to synchronize clocks had been proposed by the French physicist and clock-maker Louis Bréguet as early as 1857. For a brilliant study of the tech-

nological context of clock synchronization in the epoch of Poincaré and Einstein, see the book by Peter Galison, *Einstein's Clocks, Poincaré's Maps, Empires of Time*, New York, Norton, 2003. However, I think that knowing this background is about as relevant as knowing that in Newton's time other people had seen apples fall! Newton's genius had been to see a universal attraction reflected in the simple fall of an apple. Similarly, Einstein's genius was to profoundly modify our understanding of time, in a manner suggested by the problem of synchronization of clocks in motion. As discussed in the text, Poincaré, when faced with the same problem, continued to think of time in terms of Newton's absolute time.

[23]We take some liberties with the exact content of Einstein's article, while respecting the logical structure that supports it.

[24]Here are some hints for the enterprising reader who would like to rederive for himself the equations connecting the coordinates (x, y, z, t) in a reference frame at rest to coordinates (x', y', z', t') in a reference frame moving at a speed v along the x axis. We denote the speed of light by the letter c. For reasons of homogeneity and symmetry, we deduce that the needed relations should have the form $t' = at - bx, x' = A(x - vt), y' = By, z' = Bz$, where the coefficients a, b, A, B are functions to be determined in terms of v and c. We then note that a ray of light moving at speed c in the reference frame at rest, such that $x^2 + y^2 + z^2 - c^2 t^2 = 0$, moves also at the speed c in the moving reference frame : $(x')^2 + (y')^2 + (z')^2 - c^2 (t')^2 = 0$. Impose the symmetries under the interchange of east and west, and under the interchange between the two frames (under which, for example $B(v) = B(-v) = 1/B(v)$). Having thus obtained expressions for the coefficients a, b, A, B, verify that the combination $s^2 = x^2 + y^2 + z^2 - c^2 t^2$ is invariant under a change of reference frame (even when it is nonzero).

[25]These are the equations called (by Poincaré) Lorentz transformations. They had first been written (up to an overall factor) by the German Woldemar Voigt in 1887, then (in a more accurate form) by the Dutchman Lorentz in 1895, and finally in their exact form by the Englishman Joseph Larmor in 1900. The exact form of these equations was subsequently re-obtained by Lorentz (who did not know the work of Voigt and Larmor) in 1904. Certain properties of these equations were studied in detail by Poincaré in June 1905. Poincaré only knew Lorentz's work of 1895 and 1904, and introduced the term *Lorentz transformation*. As for Einstein, he only knew Lorentz's 1895 work, which did not contain the correct form of these equations. Independently of the question of the physical interpretation of these equations (which is completely different between Einstein and his predecessors), Einstein was the first to derive these equations in a purely *kinematical* way, through a fundamental re-analysis of the concepts of space and time.

[26]H. Poincaré, lecture at the International Congress of the Arts and Sciences, Saint Louis, Missouri, United States, September 24, 1904; published at the end of 1904, and reprinted in his remarkable popular-level book, *The Value of Science*, New York, Science Press, 1907. In all probability, Einstein had not read this lecture of Poincaré's, which foreshadowed many aspects of the theory of relativity.

[27]P. Galison, *op cit.*

[28]These courses from 1906–1907 were published in 1953:"Les limites de la loi de Newton" ("The limits of Newton's law"), Bulletin Astronomique, t. XVII, Fasc. 2, pp. 121–269.

[29]H. Poincaré, "La dynamique de l'électron" ("The Dynamics of the Electron"), *Revue générale des sciences pures et appliquées*, t. 19, pp. 386–402 (1908).

[30]In equations, $\tau = kt' = t - k^2 v(x - vt)/c^2$, where $k = 1/\sqrt{1 - v^2/c^2}$. Here, t and x are the coordinates of a reference frame "at rest," c designates the speed of light, τ is the temporal

variable of the moving observers A and B defined by Poincaré, and t' is the time seen within the reference frame which moves together with A and B, according to Einstein's definition.

[31] We note in passing that when Poincaré said that the watch of an observer in motion "runs late with respect to the other" ("*retarde sur l'autre*"), he is alluding to the fixed temporal shift between two watches in motion associated to the term linear in $x-vt$ in the synchronized "local time" that he defines: $\tau = t - k^2 v(x-vt)/c^2$. This formula, used implicitly by Poincaré, is such that the difference between two successive "local times" τ (considered at some moving location, i.e., for some fixed value of $x - vt$) is equal to the difference between the two corresponding absolute times, $\Delta \tau = \Delta t$. Poincaré never speaks of the accumulated retardation of a moving clock that returns to its point of departure, this accumulated retardation being entirely due to the supplementary factor $k = 1/\sqrt{(1 - v^2/c^2)} > 1$ in the time t' which Einstein used, which is related to the τ of Poincaré by $kt' = \tau = t - k^2 v(x - vt)/c^2$.

[32] A series of popular books which he built around the character of Mr. Tompkins.

[33] In fact, many years later, Einstein's general theory of relativity would show that the effect of gravitation, far from being negligible, is comparable to the effect of high velocity, and working in the opposite direction. The two effects essentially cancel each other out, with the result that clocks placed on a rotating, fluid earth, deformed by its rotation, tick at the same rate, whatever their positions.

[34] All the more so in that it seemed self-contradictory. Indeed, if one considers two twins in relative motion at constant speed with respect to each other, *each* of the twins will observe a *slowing* of time for the other twin compared to their own. This seems absurd. The case of two twins in relative motion *at a speed forever constant and in a straight line* does not allow one to notice, in the end, a difference of age between the two twins reunited at relative rest. To make such a test, one must, as we have always supposed, consider an asymmetrical case where one of the two twins moves with a speed whose *magnitude and/or direction varies with time*.

Chapter 2: The World's Checkerboard

[1] The following reconstruction of Einstein's visit to Paris is founded on Michel Biezunski's book, *Einstein à Paris*; see the bibliography.

[2] For a recent account of the rich life of the most famous daughter of the inventor of the automatic sewing machine, see Sylvia Kahan, *Music's Modern Muse: A Life of Winnaretta Singer, Princesse de Polignac*, Rochester, University of Rochester Press, 2003.

[3] It is regrettable that this is no longer the case. Modern newspapers and magazines (whether printed or televised) may sometimes like to make allusions to recent scientific results, but the love of the new, the sensational, and of the potentially dangerous wins out over the desire to think about the ultimate philosophical impact of science.

[4] Henri Bergson, *Durée et simultanéité. À propos de la théorie d'Einstein*, Paris, Félix Alcan, 1922. Seventh edition of the Presses universitaires de France, Paris, 1968. Available in English as *Duration and Simultaneity*, Robin Durie (ed.), Manchester, Clinamen Press, 1999.

[5] For a recent republication, see *La Pensée*, number 210, February 1980, pp. 12–29, preceded by an introduction by Michel Paty, pp. 3–11.

[6] See the comments added by the editors to the seventh edition of *Durée et simultanéité*, *op. cit.* (1968) to explain the relevance of a new edition of this book.

[7] Marcel Proust, *Lettres (1879–1922)*, selected and annotated by Françoise Leriche, Plon, 2004. Letter 572, pp. 1052–1054. I would like to thank Jean Osty for helping me discover

this letter, and Thierry Thomas for having pointed out the preparatory manuscript of *À l'ombre des jeunes filles en fleurs* cited on p. 34.

[8]This character plays a key role throughout Proust's masterpiece. In the first part of *In Search of Lost Time*, the narrator is fascinated by everything that the Princesse de Guermantes and her cousin, the Duchesse de Guermantes, symbolize: beauty, intellectual superiority, and the French nobility who occupied the very highest rank of society during the Belle Epoque. Furthermore, the work's climax consists of a description of a party given by the Prince and Princesse de Guermantes, during which the narrator has his revelation concerning *Time Regained*.

[9]Céleste Albaret, *Monsieur Proust, Souvenirs recueillis par Georges Belmont*, Paris, Éditions Robert Laffont, 1973.

[10]If one fixes each point in space by means of three orthogonal coordinates x, y, z (length, width, and height), the distance D between two points with respective coordinates (x, y, z) and $(x + \Delta x, y + \Delta y, z + \Delta z)$ is given by $D^2 = (\Delta x)^2 + (\Delta y)^2 + (\Delta z)^2$.

[11]In a similar vein, Proust, speaking about the church at Combray, writes: "Everything combined to make of it [...] an edifice occupying, if one might say, a space of four dimensions, the fourth being that of time."

[12]For a similar viewpoint, see Chapter 5 of Brian Greene, *The Fabric of the Cosmos*, New York, Alfred A. Knopf, 2004.

[13]Immanuel Kant, *Critique of Pure Reason*, translation by Norman Kemp Smith and Howard Caygill, New York, Palgrave Macmillan, 2003. For a penetrating philosophical analysis of the notion of reality, and a deep dialogue with the thought of Kant, see Martin Heidegger, *What Is a Thing?*, translated by W. B. Barton Jr. and Vera Deutsch, Chicago, Henry Regnery Company, 1967.

[14]We suppose here that the global properties of Euclidean space are identified with those of triplets of real numbers (x, y, z).

[15]Mathematically, if one fixes the space-time points by the four coordinates (x, y, z, t), where x denotes length, y width, z height, and t date, the squared interval S^2 between the events (x, y, z, t) and $(x + \Delta x, y + \Delta y, z + \Delta z, t + \Delta t)$ is given by the formula $S^2 = D^2 - c^2 T^2 = (\Delta x)^2 + (\Delta y)^2 + (\Delta z)^2 - c^2(\Delta t)^2$. Here D denotes the spatial distance between the spatial projections of the two events $(D^2 = (\Delta x)^2 + (\Delta y)^2 + (\Delta z)^2)$, $T = \Delta t$ the time interval separating their temporal projections, and c the speed of light. The product cT has dimensions of length, and expresses the distance that light travels during the interval T. Be aware of the fact that, despite the notation for the squared interval as being formally the square of a quantity S, the squared interval S^2 is not necessarily positive.

[16]In a system of units where the speed of light is numerically equal to 1.

[17]Which holds for a triangle whose three sides are all lying in the direction of time.

[18]For more detailed discussions of the concept of time in physics, see Paul Davies, *About Time*, New York, Touchstone, Simon & Schuster, 1996, and Brian Greene, *The Fabric of the Cosmos*, New York, Alfred A. Knopf, 2004.

[19]I thank John Stachel for attracting my attention to this reference.

[20]The German expression used by Einstein is "*gläubige Physiker*," which is often translated as "believing physicists." Nevertheless, all the philosophical context of Einstein's thought shows that one must not understand the word "believing" in the sense of a traditional religious belief, but rather in the sense of a deep belief in the rationality of the universe. Because of this, it

seems to us more appropriate to translate *"gläubige Physiker"* to "physicists in the soul" or to "convinced physicists."

[21] See references in the notes for the first chapter.

[22] For more details concerning the differing approaches of Poincaré and Einstein, see the books by Abraham Pais and Michel Paty cited in the general bibliography. See also the recent article "Poincaré, Relativity, Billiards, and Symmetry" by Thibault Damour, available on the electronic archives at hep-th/0501168 (see Select Bibliography).

[23] I thank David Gross for a clarifying discussion on this point.

[24] *L'espace et le temps*, lecture of May 4, 1912 at the University of London, published (with others) in Poincaré's last book: *Dernières pensées*, Paris, Flammarion, 1913. English translation by John W. Bolduc, *Mathematics and Science: Last Essays. Dernières pensées*, New York, Dover Publications, 1963.

[25] In fact, it seems that Poincaré never cited Einstein's work on relativity. He seemed at any rate to still not know of its existence in 1908. His attention was only drawn to this work in 1909 (probably on the occasion of the lectures given by Poincaré at Göttingen, and in any case through a letter from Gösta Mittag-Leffler).

[26] More precisely, the energy E contained within the mass m must be considered in the reference frame where the body is at rest.

[27] As some of Poincaré's flatterers have pushed their arguments to the point of claiming that Poincaré may have obtained the relation $E = mc^2$ in all generality before Einstein, we cite, as an example of his intuitive resistance to the equivalence between mass and energy, a sentence written by Poincaré in 1908 (three years after Einstein's work, which was probably unknown to Poincaré) in an article entitled "Electron Dynamics." Poincaré speaks of the recoil experienced by a material body that emits an electromagnetic ray in a privileged direction, and contrasts this to the recoil experienced by a cannon that emits a material projectile: "The cannon recoils, because the projectile on which it acts, reacts against it. But here, it is no longer the same. What we have sent into the distance [that is, an electromagnetic ray,] is no longer a material projectile: it is energy, and energy has no mass ... "

Chapter 3: Elastic Space-Time

[1] We speak here of the prediction made by the final version of Einstein's general theory of relativity, obtained in November 1915. As early as 1907, Einstein had understood that in any generalization of the theory of relativity allowing for gravity, there must be a gravitational influence on the propagation of light. In 1911, he obtained a provisional result suggesting that the Sun would deflect any rays passing its border by 0.875 arc-seconds. Happily for him, the first world war made it impossible to test this incomplete prediction during the eclipse of August 21, 1914.

[2] In fact, Einstein had renounced his German citizenship when he was 16 years old, and had taken up Swiss citizenship. However, he was a member of the Prussian Academy of Sciences, and Germany believed that this gave him German nationality once again. Einstein later became an American citizen as well, but retained Swiss citizenship for the rest of his life.

[3] The supposed falling body experiment made by Galileo from the Leaning Tower of Pisa is a myth, even though it sums up well the essence of Galilean innovation.

[4] Galileo Galilei, *Dialogues Concerning Two New Sciences*, translated by Henry Crew and Alfonso di Salvio, New York, Macmillan, 1914.

[5] One has added the prefix "hyper" to the word surface to indicate that this ensemble of points has one dimension less than the ambient space in which it is traced. As it is traced in a space-time with four dimensions, this means that it has three internal dimensions (while a surface in ordinary three-dimensional space has only two dimensions). The mathematical name designating what we here call an hourglass is "hyperboloid."

[6] We are here alluding to what is called, in mathematics, a general ellipsoid.

[7] In precise mathematical terms, this is a general hyperboloid.

[8] In fact, it was subsequently understood that the principle of general relativity is not, physically speaking, a generalization of the principle of (special) relativity. The principle of special relativity is a *symmetry* principle of the structure of space-time, which asserts that physics is *the same* for a particular class of reference frames, and thus that certain corresponding phenomena play out in exactly the same way in different reference frames (related by "active" transformations). On the other hand, the principle of general relativity is a *principle of indifference*: phenomena may not (in general) play out in the same fashion in different coordinate systems, but none of the (global) coordinate systems has a privileged status with respect to the others.

[9] In relation to what we have remarked previously, the Pythagoras-Einstein theorem in a deformed space-time, tiled by four arbitrary coordinates x_0, x_1, x_2, and x_3, states that the squared interval between two points infinitesimally close together (with coordinates x_0, x_1, x_2, x_3 and $x_0 + dx_0$, $x_1 + dx_1$, $x_2 + dx_2$, $x_3 + dx_3$) is a sum of terms proportional to the squares and double products of the (infinitesimal) coordinate differences: dx_0, dx_1, dx_2, dx_3. There are *ten* terms in this sum, since there are four squares, dx_0^2, dx_1^2, dx_2^2, dx_3^2 and six double products $2dx_0dx_1$, $2dx_0dx_2$, $2dx_0dx_3$, $2dx_1dx_2$, $2dx_1dx_3$, and $2dx_2dx_3$. The coefficients of the four squares are denoted, respectively, by g_{00}, g_{11}, g_{22}, and g_{33}, while the coefficients of the double products are denoted g_{01}, g_{02}, g_{03}, g_{12}, g_{13}, and g_{23}. If we call ds^2 the infinitesimal squared interval between the two points considered, we can write the Pythagoras-Einstein theorem in the form $ds^2 = \sum g_{\mu\nu}dx_\mu dx_\nu$, where each index, μ or ν, takes the four values 0, 1, 2, and 3 and the sign $\sum$ indicates that one sums, independently, over the two indices μ and ν. Einstein simplified this notation (due to Riemann) by remarking that it was unnecessary to write the symbol $\sum$, since it sufficed to demand implicitly that one must sum over every repeated index (here μ and ν). Einstein always wrote the indices μ and ν as subscripts on the coordinates x. Today they are written as *superscripts* (even though this sometimes can be confused with exponents). Thus, we write finally:

$$ds^2 = g_{\mu\nu}(x^\lambda)dx^\mu dx^\nu,$$

where we have indicated that the ten coefficients $g_{\mu\nu}$ are functions of the four coordinates x^λ.

[10] In fact, the mathematical term "tensor" finds its origin in the physical application of this object, describing the tensions within a continuous medium.

[11] One may show that the deformation tensor is mathematically constructed through various spatial derivatives of the displacement vector of the gelatin. The displacement vector is the ensemble of small arrows joining the initial, nonperturbed, position of a material point in the gelatin to its final, perturbed position.

[12] In this relation, it can be shown, for homogeneous and isotropic media, that the object $\hat{x}$ is a little more complicated than a simple numerical coefficient of proportionality. It is constructed from two distinct numerical coefficients, known as Lamé elasticity coefficients.

[13] The components $g_{\mu\nu}$ will take the value +1 when the indices μ and ν are equal to each other and correspond to the square of a spatial coordinate difference, that is, when $\mu = \nu = $

1 or 2 or 3. The component g_{00} corresponding to the square of the temporal coordinate difference will take the value -1 if one uses $x^0 = ct$ as time coordinate (or $g_{tt} = -c^2$ if one directly uses the time t as temporal coordinate). Finally the six remaining components corresponding to the double products, the components $g_{\mu\nu}$ when μ is different from ν, will all be zero.

[14] In the usual notation, this tensor is denoted $T^{\mu\nu}$, where the indices μ and ν correspond to the x^μ coordinates used, with $\mu = 0, 1, 2, 3$. The component corresponding to the "square of time," that is T^{00}, measures the density of mass-energy, while the purely spatial components T^{ij}, where i and j take the values 1, 2, 3, correspond precisely to the stress tensor of elastic media.

[15] Nevertheless, as he had not yet obtained the definitive formulation of his theory of gravitation, he predicted, in Prague, a deflection of light smaller by a factor of two than his final result. Specifically, he obtained 0.875 arc-seconds (a value that one can obtain in the Newtonian theory of gravitation if one considers light to be composed of corpuscles), in place of the value of 1.75 arc-seconds that he would find in November 1915.

[16] We note that Einstein never used the expression, used in this book, of law of space-time elasticity. Nevertheless, we believe we are not betraying the central idea of his theory, but rather clarifying it, by using this image.

[17] This is called the Ricci tensor.

[18] Hannes Alfven, *Cosmology: Myth or Science?*, in *Einstein, A Centenary Volume*, A. P. French (ed.), Cambridge, Harvard University Press, 1980. This passage is translated from the French version, *Einstein, Le Livre du Centenaire*, prepared by G. Delacôte and J. Souchon-Royer, Paris, Hier et Demain, 1979, p. 83. Cited by Michel Biezunski, *Einstein à Paris, op. cit.*

[19] The tidal tensor, also called the gravity gradient, is the mathematical object defined by taking two successive spatial derivatives of the Newtonian gravitational potential. The tensor **R** is a more complicated object, derived from **g**, and of the form $\mathbf{R}(\mathbf{g}) = \mathbf{g}^{-1}\mathbf{ddg} + \mathbf{g}^{-1}\mathbf{g}^{-1}\mathbf{dgdg}$, where **g** designates the ten components of the metric tensor $g_{\mu\nu}$, $\mathbf{g}^{-1} = g^{\mu\nu}$ the inverse matrix of $g_{\mu\nu}$, and **d** is a spatio-temporal gradient, that is, a partial derivative with respect to the four space-time coordinates x^μ. The mathematical object used by Einstein, and denoted $\mathbf{D(g)}$ in the text, has exactly the same structure as $\mathbf{R(g)}$, that is, it contains (linearly) the second derivatives of **g** and it is quadratically nonlinear in the first derivatives of **g**.

[20] For the aficionados who are not afraid to see some equations, we note that the Riemann tensor has four independent indices, $\mathbf{R} = R^\alpha{}_{\beta\mu\nu}$, and that one successively derives, by summation over certain indices, the Ricci tensor $R_{\mu\nu} = R^\alpha{}_{\mu\alpha\nu}$, and then the Einstein tensor $D_{\mu\nu} = R_{\mu\nu} - (1/2)Rg_{\mu\nu}$, where $R = g^{\mu\nu}R_{\mu\nu}$. Einstein's equations are thus finally $D_{\mu\nu} = R_{\mu\nu} - (1/2)Rg_{\mu\nu} = \varkappa T_{\mu\nu}$, where $T_{\mu\nu}$ is the stress-energy tensor. There is no standard notation for the Einstein tensor (denoted here by $D_{\mu\nu}$). The most common notations are $G_{\mu\nu}, S_{\mu\nu}$, or $E_{\mu\nu}$.

Chapter 4: Einstein's World-Game

[1] The left-hand side of the equation $\mathbf{D}' = \varkappa\mathbf{T}$ that he proposed on November 11th differed from the correct final result in that $\mathbf{D}'$ was only the Ricci tensor, instead of the Einstein tensor, which differs from that of Ricci by the supplementary term $-(1/2)R\mathbf{g}$. Einstein would write the final form of **D** on November 25th. It was believed for a long time (and some authors of books about Einstein continue to believe) that the mathematician Hilbert had realized, as

early as November 20th, a full five days before Einstein's final article, the necessity of adding the supplementary term $-(1/2)R\mathbf{g}$ in the equation written on the 11th by Einstein. However, the recent discovery of Hilbert's original, corrected proofs has shown that he extensively modified the original version of his article, after having read Einstein's final article dating from November 25th.

[2] In ordinary space, the straightest lines are the shortest lines. However, in space-time, because of the minus sign which affects the temporal directions, the straightest possible lines (directed "in time") are found to be the longest.

[3] To use an expression that Einstein will apply, some years later, to a new idea of Louis de Broglie.

[4] We note in passing that Poincaré had realized, as early as June 1905, that any "relativistic" theory of gravity would predict phenomena connected to the propagation of gravity at the speed of light (which he baptized as *ondes gravifiques*, or gravity waves). He also predicted that, probably, these *ondes gravifiques* would drain the energy of their source. In 1908, he suggested an observable phenomenon tied to this energy loss: the secular acceleration of the orbital frequency of a planetary system. It is remarkable that it is by this phenomenon (observed in the binary pulsar PSR 1913 + 16) that the reality of gravitational waves was confirmed in the 1980s. We note nevertheless that Poincaré's reasoning (inspired by previous considerations of Laplace and Lorentz) was purely qualitative. In contrast to Einstein, Poincaré never proposed a specific theory of relativistic gravitation. He lacked some essential tools, which for Einstein were the principle of equivalence and the principle of general relativity.

[5] By H. A. Lorentz (of the Lorentz transformations) and J. Droste. An equivalent result was later obtained (in 1938) by Einstein, L. Infeld, and B. Hoffmann by a new method which turned out to be important for describing the motion of neutron stars or black holes.

[6] In work by T. Damour, completing earlier work made in collaboration with N. Deruelle, as well as with L. Bel and J. Martin.

[7] George Gamow, *Mr. Tompkins in Paperback*, Cambridge, Cambridge University Press, 1982, Chapter 3.

[8] We here neglect the fact that the deformation wave oscillates at a relatively high frequency (on the order of 100 Hertz for the sources being sought by LIGO and VIRGO).

[9] But only in the absence of the supplementary term, called the cosmological constant, that Einstein introduced in his foundational article on cosmology in 1917. In fact, one of the arguments used by Einstein to introduce this supplementary term is to keep Minkowski space-time from being a solution, in the absence of matter.

[10] We prefer the expression *curved in time* to the expression *expanding* (or *contracting*) to avoid the clandestine reintroduction of a notion of *temporal flow*. See the discussion below.

[11] See, for example: Joseph Silk, *The Big Bang*, New York, W. H. Freeman, 2001, and *A Short History of the Universe*, New York, W. H. Freeman, 1997; and Brian Greene, *The Fabric of the Cosmos*, op. cit.

[12] We use the term *big crunch* for a space-time border which would be a border along the top, in the arbitrary convention where the Big Bang in the usual sense is considered as a border along the bottom of space-time. In other words, if one (mentally) cuts space-time into slices along a "cosmic time" which measures the height above the Big Bang (that is, a cosmic time which is zero at the Big Bang and has a positive value in the part of space-time that we inhabit), a big crunch is the temporal inverse of a big bang.

[13]That is to say, conceivable. We note that possible is not the same as probable, even if, in the quantum theory, anything possible is mandatory, in other words realized with a nonzero *existence amplitude* (usually called a *probability amplitude*). Everything we know indicates that the part of space-time to which we have access is in a state with a privileged *temporal orientation*, reflected in the temporal stratification of many (cosmological, astrophysical, electromagnetic, thermodynamic, etc.) structures.

[14]One must be careful not to confuse *temporal orientation* (or *arrow of time*) with *temporal flow*. For example, a block of gelatin could, for example, because of the sedimentation of some of its constituents during the refrigeration which formed it, be *stratified*, denser below and less dense above (with a continuous variation of density from bottom to top). But this privileged arrow does not imply that something is moving upwards. Similarly, our space-time is not homogeneous but it is *stratified*. The privileged strata are space-like: they are aligned with *positive* squared intervals, but there is nothing that corresponds to the idea of a stratum of the present which would "move" towards the future, like a projector successively lighting the strata of constant density of space-time.

[15]It should be noted that the thermodynamic arrow of time (the direction of time with respect to which entropy grows) is what determines the sensation of the passage of time, through the irreversibility of the process of memorization in the neuronal structures which give rise to the phenomenon of consciousness. In a cosmological model of the type considered in the text, the thermodynamic arrow of time would not be well-defined in certain transitional regions, where the entropy would pass through some maximal values.

[16]For more on Gödel's views on time, and on his friendship with Einstein, see Palle Yourgrau, *A World without Time: The Forgotten Legacy of Gödel and Einstein*, Cambridge, MA, Basic Books, 2005.

[17]We here consider the deformation of the spatial geometry: the geometry of slices of space-time considered at a given instant.

[18]We shall assume here that the generic singularities of space-time are all locally similar to a cosmological singularity (extended on a space-like or, at most, a light-like, hyper-surface). This hypothesis (a simplification for us) is founded on some definite results but remains essentially an open conjecture in nonquantum general relativity.

[19]For an introduction to the astrophysics of black holes, and to their history, see Jean-Pierre Luminet, *Black Holes*, Cambridge, Cambridge University Press, 1992; and Werner Israel, *Dark Stars: The Evolution of an Idea*, in *300 Years of Gravitation*, edited by S. W. Hawking and W. Israel, Cambridge, Cambridge University Press, 1987.

[20]Mathematically, the chronogeometry represented in this diagram (apart from the collapse of the star) is that of a Schwarzschild black hole, corresponding to a solution of Einstein's equations obtained by Karl Schwarzschild, and independently by Johannes Droste, in 1916. For the aficionados, here is the mathematical form of the infinitesimal squared interval of this chronogeometry: $ds^2 = -Ac^2 dt^2 + dr^2/A + r^2(d\vartheta^2 + (\sin\vartheta)^2 (d\varphi)^2)$, where r is a radial coordinate, $A = 1 - 2GM/(c^2 r)$, and where ϑ designates the latitude (starting from the north pole) and φ the longitude of a directional sphere. The horizon (outside of the star) of the Schwarzschild black hole is the "light cylinder" $r = 2GM/c^2$.

[21]The energy, momentum, and angular momentum of a black hole are defined by the formalism introduced by Richard Arnowitt, Stanley Deser, and Charles Misner.

[22]In work dating from 1971 where they showed the existence of a fundamental irreversibility in the physics of black holes.

[23] The notion of the entropy of a black hole was introduced by Jacob Bekenstein in 1973.

[24] The notion of black hole temperature was introduced by Stephen Hawking in 1974, in a calculation where he discovered the remarkable phenomenon of quantum evaporation of black holes.

[25] The notion of black hole surface resistivity was introduced, independently, by Thibault Damour and Roman Znajek in 1978.

[26] The notion of black hole surface viscosity was introduced by Thibault Damour in 1979.

Chapter 5: Light and Energy in Grains

[1] The explanations given here concerning Ernest Solvay, and the first Solvay council, are in large part pulled from Pierre Marage and Grégoire Wallenborn (editors), *Les Conseils Solvay et les débuts de la physique moderne (The Solvay Councils and the Beginnings of Modern Physics)*, Brussels, Université libre de Bruxelles, 1995.

[2] The International Solvay Institute, based in Brussels, continues this tradition with great success today, and with the continued support of the Solvay family.

[3] Notably by Gustav Kirchhoff, Joseph Stephan, Ludwig Boltzmann, Wilhelm Wien, Friedrich Paschen, Max Planck, Otto Lummer, Ernst Pringsheim, Heinrich Rubens, and Ferdinand Kurlbaum.

[4] This law is now called the Rayleigh-Jeans law. In fact, as Abraham Pais remarked in his biography of Einstein (see Select Bibliography), this law should have been called the Rayleigh-Einstein-Jeans law because Einstein was the first to give it a complete derivation, and to understand its significance. The original work of Lord Rayleigh (1900) did not calculate the global multiplicative factor appearing in this law.

[5] In practice, we do not use the base ten logarithm, but that associated to the "natural" base $e = 2.71828$. Thus, one writes $N = e^L$.

[6] We here simplify the historical development of the link between entropy, statistics, probability, and the number of microscopic states. To be more complete, one should cite the contributions of J. C. Maxwell, Max Planck, and Josiah Willard Gibbs, and of Einstein himself. In fact, in 1905, few physicists had understood and accepted the link between entropy and probability. Einstein's first papers, before 1905, were concerned with this link and, even if they did not contain anything revolutionary, they furnished him with some very powerful intellectual tools for his future research.

[7] In fact, the notion of *the number of microscopic states* only became well defined in quantum theory. It, however, remains remarkable that an adequately improved use of pre-quantum statistics allowed Planck, and especially Einstein, to lay the foundations of quantum theory (and quantum statistics).

[8] Actually, Einstein preferred to think in terms of *probability*, and he used entropy differences to estimate relative probabilities, with a higher entropy being connected to a higher relative probability.

[9] The notation h is that introduced by Planck in 1900. According to the interesting little book *Max Planck et les Quanta*, by Jean-Claude Boudenot and Gilles Cohen-Tannoudji, Paris, Ellipses, 2001, Planck chose the letter h for *hilfe grösse*, or auxiliary magnitude. We note also that Einstein did not use the notation h in his 1905 article (nor for many years). Since he thought (somewhat correctly) that the derivation of the law of black-body radiation given

by Planck in 1900 was self-contradictory, he preferred to present his arguments in a fashion independent of Planck's reasoning.

[10] With the condition that one leaves fixed the physical size of available initial and final surfaces. For example, if the initial surface had an area of one squared centimeter, and the final surface an area of 64 squared centimeters, we could have used a checkerboard with squares each of size one millimeter squared. The initial surface would then be composed of 100 elementary squares, and the final surface composed of 6,400 squares.

[11] As Olivier Darrigol has remarked (in his contribution to the collection *Einstein aujourd'hui*; see Select Bibliography), Planck, a scientist of great integrity, acknowledged in his Nobel lecture of 1920 that science was indebted to Einstein for having considered the quantization of the energies of an oscillator as a physical reality (rather than as a purely formal size of "elementary domains of probability").

[12] This is essentially due to the equipartition theorem of classical statistics, which states that each mechanical degree of freedom has an average thermal kinetic energy of $kT/2$, where k is Boltzmann's constant and T the absolute temperature. One must also incorporate the fact that, for an oscillator as for a spring, the average potential energy is equal to the average kinetic energy. This gives a specific heat of $3k$ per atom.

[13] Some years later, it was suggested by Planck (in 1911), then demonstrated by the new quantum theory of Werner Heisenberg and Erwin Schrödinger (1925–1926), that even in the absence of any external agitating force (that is, at zero absolute temperature) the fundamental state, with the lowest possible energy, of an oscillator did not have zero energy, but an energy equal to half of the jump to its first level of excitation: $hf/2$. In other words, the vibrational energy of a quantum oscillator could only take the values $hf/2, 3hf/2, 5hf/2, 7hf/2$, and so on.

[14] It was subsequently remarked that Einstein's result was not in very good agreement with the data at very low temperatures. However, a refinement of Einstein's calculation due to Peter Debye, which retained the essential idea (the quantization of the oscillator energy), agreed excellently with experiment.

[15] Apart from a new proof, more general than earlier ones, that the law of a black body must be Planck's law.

[16] The term *photon* was used (in writing) for the first time in 1926 in an article by the American physical chemist Gilbert Lewis. However, it is clear that the concept was introduced by Einstein in this 1916 article.

[17] We recall that Alfred Kastler, as a young student, was stimulated by hearing Einstein in person when he had visited Paris in the spring of 1922. It seems, however, that Einstein had spoken there only of relativity.

[18] In fact, Bose introduced, without really understanding it, a new method of applying the statistical method to a gas of identical quantum particles which were *indistinguishable*. This new aspect of Bose's work was enlightened by the work of Einstein, particularly through questions raised by Paul Ehrenfest.

[19] More precisely, it is the square of the fluctuation which decomposes into the sum of two separate terms.

[20] According to Wolfgang Pauli, Einstein proposed, at a physics conference held at Innsbruck from the 21st to the 24th of September 1924, the search for the existence of interference and diffraction for molecular beams. The letters written by Einstein in December 1924 to Langevin and Lorentz, which conveyed the enthusiasm he had recently experienced after

reading the thesis of Louis de Broglie (defended in Paris on the 25th of November 1924), suggest that Einstein became aware only in December 1924 of the ideas (dating from 1923) of de Broglie on the light-matter similitude. Einstein then recognized de Broglie's precedence, and made himself a champion of this similarity, without insisting on the fact that he had independently found weighty arguments which suggested it. We note nevertheless that Abraham Pais, in his biography, cites a letter that de Broglie wrote to him in 1978, in which he suggests that Langevin had communicated a copy of his thesis to Einstein in the spring of 1924.

[21] The second relation (due, in the general case, to de Broglie) is a natural consequence of the first relation (of Planck-Einstein) if one applies the ideas of special relativity (due to Einstein). We note that the relation $\lambda = cT$, which applies to light, is no longer applicable to the case of a "massive" material particle.

Chapter 6: Confronting the Sphinx

[1] We are here inspired by the memoirs collected, much later, by Werner Heisenberg in his remarkable book *Physics and Beyond*, translated by A. J. Pomerans, London, Allen and Unwin, 1971.

[2] The book *Space, Time, Matter* by the mathematician Hermann Weyl, was one of the first books written on the theory of general relativity. The first edition is dated 1918.

[3] We recall that the possible energies, in quantum theory, for the states of an atom only take discontinuous values E_0, E_1, E_2, etc. The coefficient which Einstein associated to the quantum transition between the state of energy E_m and the state with (lower) energy E_n is denoted A_{nm}. Here m and n are indices which take the values 0, 1, 2, etc. If f_{nm} designates the frequency of light emitted during the transition between "the state m" and "the state n" (as we shall say for brevity), the energy of the quantum of light emitted is $E = hf_{nm} = E_m - E_n$, and its momentum takes the value $p = hf_{nm}/c$.

[4] The amplitude a_{nm} associated to the transition between the state m and the state n is a complex number ($a_{nm} = x_{nm} + iy_{nm}$, where $i = \sqrt{(-1)}$), of which the squared modulus ($|a_{nm}|^2 = |x_{nm}|^2 + |y_{nm}|^2$) is proportional to Einstein's coefficient A_{nm} associated to the same transition.

[5] Like Heisenberg in his first article, we here consider for simplicity an atom with only one electron.

[6] Born quickly realized that the table a_{nm} of (complex) amplitudes considered by Heisenberg could be identified with what the mathematicians called a matrix, since the calculational rules introduced by Heisenberg, for physical reasons, were found to be the same as the rules for matrix calculations. We note, however, that the table of transition amplitudes a_{nm} is infinite, in general.

[7] To repeat an expression used by Einstein on December 25, 1925, in a letter to Besso.

[8] See Chapter 5.

[9] See Chapter V of the memoirs of Heisenberg cited above.

[10] These are the table of values $f_{nm} = (E_m - E_n)/h$ and that of the values a_{nm} mentioned in the notes above.

[11] More precisely, $\mathcal{A}$ is a complex function ($\mathcal{A} = \mathcal{A}_1 + i\mathcal{A}_2$). This wave amplitude is often denoted, following Schrödinger, by the Greek letter psi, ψ.

[12] For historical references on Einstein's ghost field (*Gespensterfeld*) and on its influence on the probabilistic interpretation of the wave amplitude $\mathcal{A}$, see the biographies of Abraham Pais

on Einstein (see Select Bibliography) and of Bohr (*Niels Bohr's Times*, Oxford, Clarendon Press, 1991).

[13] $\mathcal{A}$ being a complex number, $\mathcal{A} = \mathcal{A}_1 + i\mathcal{A}_2$, the square we speak of here is the squared modulus of $\mathcal{A}$: $|\mathcal{A}|^2 = (\mathcal{A}_1)^2 + (\mathcal{A}_2)^2$.

[14] See Chapter VI of his book: *Physics and Beyond, op. cit.*

[15] Also known as *indeterminacy relations* or *dispersion relations*.

[16] Recall that the (relativistic) momentum of a particle is given by $p = mv/\sqrt{(1 - v^2/c^2)}$, where m is the particle's mass (at rest), and v its speed.

[17] Depending on the precise technical definition of uncertainty, the minimum of their product may differ from h by a numerical factor.

[18] In the sense that certain physicists followed Einstein in his doubts concerning the definitive and/or complete character of the quantum theory, while the majority rallied around the "Copenhagen interpretation."

[19] For an introduction to the modern approach to the problem of motion of gravitationally condensed objects, see T. Damour, "The Problem of Motion in Newtonian and Einsteinian Gravity," in *300 Years of Gravitation*, edited by S. W. Hawking and W. Israel, Cambridge, Cambridge University Press, 1987, Chapter 6, pp. 128–198.

[20] We note that there is nothing "incorrect" in Bohr's response, and that moreover it would not be "incorrect" to say that recent experiments on the EPR system have "vindicated" Bohr. The author, however, thinks that Einstein's approach, translating conceptual questions into thought experiments (which were subsequently realized) reflected a better sense of physics than that of an *a priori* rejection of any need for experimental verification through a quasi-religious belief in the metaphysically fuzzy concept of complementarity.

Chapter 7: Einstein's Legacy

[1] The content of this lecture is known to us through notes taken by John A. Wheeler during the seminar, and by the memories reported by some of the attendees. See pp. 201–211 of the book edited by Peter C. Aichelburg and Roman U. Sexl, *Albert Einstein, His Influence on Physics, Philosophy, and Politics*, Braunschweig/Wiesbaden, Vieweg, 1979.

[2] Recall that it is Einstein himself who introduced probability into quantum theory in the 1916 article where he described precisely this transition process between atomic levels under the influence of electromagnetic radiation.

[3] I thank Charles W. Misner for having confirmed to me the presence of Hugh Everett at this lecture. For a detailed biography of Hugh Everett III, see the text by Eugene Shikhovtsev (edited by Kenneth Ford) on Max Tegmark's website: http://space.mit.edu/home/tegmark/everett/index.html. We have pulled the greater part of the facts concerning Everett cited in the text from this biography.

[4] I do not know if Everett had explicitly heard this phrase. He could have heard of its existence through John Wheeler, who must have known it. This phrase figures prominently in the book of John Archibald Wheeler and Wojciech Hubert Zurek, *Quantum Theory and Measurement*, Princeton, Princeton University Press, 1983.

[5] See the Everett biography by Eugene Shikhovtsev (edited by Kenneth Ford), *op. cit.*

[6] Bryce DeWitt, *The Global Approach to Quantum Field Theory*, Volume 1, page 144, Oxford, Clarendon Press, 2003.

[7]To be more precise, we must consider all of the stable elementary particles of the system (electrons, quarks) and include as well a description of the various interaction fields (electromagnetic, weak and strong nuclear, and gravitational).

[8]In other words $q = (x_1, y_1, z_1; x_2, y_2, z_2; \ldots; x_N, y_N, z_N)$. The amplitude $\mathcal{A}$ is a complex function of the time t (at which the configuration is considered) and of the $3N$ real variables q.

[9]Thibault Damour and Jean-Claude Carrière, *Entretiens sur la multitude du monde*, Paris, Éditions Odile Jacob, 2002.

[10]An ordinary photographic image is an imperfect representation since it projects a three-dimensional configuration onto a flat, two-dimensional film. The reader must imagine that either we are speaking of three-dimensional photographs or of two-dimensional holograms containing all of the spatial information of the configuration.

[11]More precisely, the frequency f with which the hue of a physical system turns on the color wheel is given by the Planck-Einstein relation ($E = hf$). That is, it takes the value $f = E/h$ where E is the total energy of the system and h is Planck's constant. This link between the energy of the system and the frequency of rotation around the circle of the complex amplitude $\mathcal{A}$ essentially constitutes the famous Schrödinger's Equation. Because of the extremely small numerical value of Planck's constant h, the frequency f is extremely large for any macroscopic energy E.

[12]Later, other physicists, notably Bryce DeWitt, would improve the proof sketched by Everett.

[13]One of the first physicists to understand the role of decoherence in quantum theory was H. Dieter Zeh (1970). The first rigorous result on decoherence, and on its role in justifying the quantum theory of measurement, is due to the Swiss mathematical physicist Klaus Hepp (1972). Decoherence is presently the object of many experimental studies (notably by the group led by the French physicist Serge Haroche). It is indeed essential to understand and master decoherence in order to envisage utilizing all the possibilities offered by quantum theory in cryptography and computation.

[14]See the stimulating book by David Deutsch, *The Fabric of Reality: The Science of Parallel Universes and Its Implications*, New York, Penguin Books, 1997.

[15]See Immanuel Kant, *Critique of Pure Reason*, New York, Palgrave Macmillan, 2003. See also the previously cited book of Martin Heidegger, *What Is a Thing?*

[16]In French, *Le Kantique du Quantique*, which is a (nontranslatable) play on the French *Le Cantique des Cantiques*, which refers to the "Song of Songs" in the Bible.

[17]Some recent experiments, due notably to the group of the physicist Serge Haroche, have permitted detailed observation of situations of the Schrödinger's Cat type for *mesoscopic* systems (intermediate between the microscopic level and the macroscopic level).

[18]See, for example, T. Damour and J.-C. Carrière, *Entretiens sur la multitude du monde*, Paris, Éditions Odile Jacob, 2002.

[19]If p equals one, we find a string, while $p = 2$ gives a membrane, $p = 3$ an elastic mollusk, etc. The case $p = 0$ describes a point-like particle. Even the case $p = -1$ exists and describes an instanton, which is an object (introduced by A. Polyakov) which only exists for a brief instant at one point in space.

[20]The fact that the curved geometry of space-time appears in string theory as a "correction" to an initial undeformed space-time is unsatisfactory. Many physicists hope that string theory

will one day satisfy a principle of generalized general relativity, where it will not be necessary to start by providing a background space-time.

[21] In this theory, the metric tensor $g_{\mu\nu}$ is no longer supposed to be symmetric in the indices μ and ν. The symmetric part of $g_{\mu\nu}$ corresponded to the usual geometry of general relativity, while the antisymmetric part was a new field. It is found that the equations written by Einstein are similar to those following from string theory, where there naturally arises at the same time a symmetric tensor and an antisymmetric tensor (the Neveu-Schwarz $B_{\mu\nu}$ field).

[22] I here allude to the "duality" between gauge theories and strings, such as has been conjectured by Alexander Polyakov, as well as by Juan Maldacena.

[23] Initiated by Igor Klebanov, and amplified by a remarkable conjecture of Juan Maldacena.

[24] Conforming to the general theme of this book, we concentrate here on Einstein's contributions. Of course, it would be even more enriching to think also about the fact that the photoelectric effect had been discovered by Heinrich Hertz in a fortuitous way in his experiments trying to establish the reality of electromagnetic waves. One should also think of all those scientists and engineers whose "divine and playful curiosity" has contributed in an essential fashion to the understanding and use of the photoelectric effect: notably Jean Perrin and Joseph John Thomson, who would discover the electron, and Philipp Lenard, who discovered the existence of the threshold frequency in the photoelectric effect.

[25] The fractional relativistic effects on the apparent frequency of clocks is on the order of 10^{-9}, which is small in absolute value, but which is all the same 10,000 times greater than the precision of atomic clocks (on the order of 10^{-13} or better).

[26] Tony Cawkell and Eugene Garfield, "Assessing Einstein's Impact on Science by Citation Analysis," in *Einstein: The First Hundred Years*, Oxford, Pergamon Press, 1980, p. 32; I am here inspired by the beginning of Chapter 7 of Albrecht Fölsing's book *Albert Einstein*; see the bibliography.

[27] This citation is extracted from the excellent little book of Helen Dukas and Banesh Hoffmann, *Albert Einstein, the Human Side: New Glimpses from His Archives*, Princeton, Princeton University Press, 1979.

[28] The word "Bern" is derived from the Swiss-German word *Bär*, which means bear.

Select Bibliography

In order to not encumber the text, we have not systematically indicated the origins of facts or cited texts. We here repair in part this omission by mentioning the principal sources on which we have relied. Most important are:

Albert Einstein, *Œuvres choisies* (in French), under the direction of Françoise Balibar, 6 volumes, Paris, Éditions du Seuil, 1989–1993. These selected works present French translations of Einstein's principal articles (and writings), completed with remarkable historical presentations due to a team of science historians (notably Françoise Balibar, Olivier Darrigol, Jean Eisenstaedt, and John Stachel). We have made particular use of Volumes 1 (*Quanta*), 2 (*Relativity I*), and 3 (*Relativity II*). The complete works of Einstein are under publication in their original language (i.e., mostly German) at the Princeton University Press. The nine volumes published up until now cover the years up to 1920. Literal English translations are also published in separate volumes.

The Einstein biographies that we have made particular use of are:

Albrecht Fölsing, *Albert Einstein*, New York, Penguin Books, 1998; English translation (by Ewold Osers) from the original German: *Albert Einstein: Eine Biographie*, Frankfurt am Main, Suhrkamp Verlag, 1993. This book is the biography of reference on Einstein.

Abraham Pais, *"Subtle Is the Lord ... The Science and the life of Albert Einstein*, Oxford, Oxford University Press, 1982. The best overall analysis of Einstein's scientific work.

Banesh Hoffmann and Helen Dukas, *Albert Einstein, Creator and Rebel*, New York, The Viking Press, 1972. A fervent biography written by one of Einstein's collaborators and Einstein's secretary.

For the reader who desires to have a rapid overview of the whole of Einstein's life and work, we recommend:

Françoise Balibar, *Einstein: Decoding the Universe*, New York, Harry N. Abrams Books, 1993. A thorough and magnificently illustrated little book.

We warmly recommend the direct reading of the writings by Einstein himself. Notably:

Albert Einstein, *Relativity: The Special and the General Theory*, translated by Robert W. Lawson, New York, Three Rivers Press, 1961.

Albert Einstein, the Human Side: New Glimpses from His Archives, selected and edited by Helen Dukas and Banesh Hoffmann, Princeton, Princeton University Press, 1979.

Albert Einstein, *The World as I See It*, New York, Carol Publishing Group, 1991.

Albert Einstein and Leopold Infeld, *The Evolution of Physics: The Growth of Ideas from Early Concepts to Relativity and Quanta*, New York, Simon and Schuster,1938.

Pierre Speziali (ed.), *Albert Einstein-Michele Besso, Correspondance 1903–1955*, German with parallel French translation and notes by Pierre Speziali, Paris, Hermann, 1972. Fascinating and touching correspondence between Einstein and his most intimate friend.

For an in-depth study of Einstein's physics as philosophical practice, see:

Michel Paty, *Einstein philosophe*, Paris, Presses universitaires de France, 1993, in French.

Einstein's visit to Paris in 1922 is the subject of:

Michel Biezunski, *Einstein à Paris*, Saint-Denis, Presses universitaires de Vincennes, 1991, also in French.

We mention in passing that the translations of Einstein's writings given in the text have often been modified (or made) by us starting in general from other translations (in French or in English). In some cases, we have been able to refer directly to the original German. For a useful collection of quotes (translated into English) of Einstein on many different topics, see Alice Calaprice, *The New Quotable Einstein*, Princeton University Press and The Hebrew University of Jerusalem, 2005. We note with regret, however, that this collection does not include the German originals of the quotes.

For the reader who wants to measure the impact of Einstein's work on the physics of today, see:

Alain Aspect et al., *Einstein aujourd'hui*, Paris, EDP-Sciences and CNRS Éditions, collection "Savoirs actuels," 2005, in French. Each chapter of this book is written by one or more experts on one aspect of Einstein's work.

Finally, the reader can have access, freely, to work in progress in relativistic and/or quantum physics by connecting to the various archives of http://www.arxiv.org/, notably the archive gr-qc (General Relativity and Quantum Cosmology). For advanced searches of (technical) works in fundamental physics, see http://www.slac.stanford.edu/spires/hep/. Let us also point out the existence of websites where one can access some of Einstein's original manuscripts: http://www.albert-einstein.org/ and http://www.alberteinstein.info/.